CREATION COLLECTION

UNIVERSE

Surveying God's Created Cosmos

CREATION COLLECTION

UNIVERSE

Surveying God's Created Cosmos

Jake Hebert with Brian Thomas, Frank Sherwin,
Timothy Clarey, and Jeffrey P. Tomkins

ICR INSTITUTE FOR CREATION RESEARCH

Wayfinders PRESS

Dallas, Texas
ICR.org

Universe
Surveying God's Created Cosmos

Contributions by Jake Hebert, Ph.D., with Brian Thomas, Ph.D., Frank Sherwin, D.Sc. (Hon.), Timothy Clarey, Ph.D., and Jeffrey P. Tomkins, Ph.D.

First Printing: May 2024

Designed by Susan Windsor

Concept by the ICR Communications team: Jayme Durant, Beth Mull, Michael Stamp, Lori Fausak, Renée Dusseau, Bethany Trimble, Rachel Brown, Melissa Marquez, and Frank Sherwin

All Scripture quotations are from the New King James Version.

ISBN (paper): 978-1-957850-57-3
ISBN (ebook): 978-1-957850-65-8
Library of Congress Control Number: 2024937822

Please visit our website for other books and resources: ICR.org

Printed in the United States of America.

Published by Wayfinders Press, an imprint of ICR Publishing Group

Cover: Near-infrared-light view of the Pillars of Creation taken by NASA's James Webb Space Telescope

TABLE OF CONTENTS

INTRODUCTION

The universe inspires awe with its countless wonders and mysteries. Its fascinating marvels have sparked centuries of exploration, from the first telescope to the first man on the moon. Even today, we don't fully understand all the incredible complexities of our vast cosmos.

Evolutionists claim that everything in this beautiful and ordered universe, including you, came from a chaotic explosion billions of years ago. This Big Bang theory is so widely held that some Christians think it was used by God at the beginning—even though much of it contradicts Scripture.

In *Universe: Surveying God's Created Cosmos*, ICR scientists examine the Big Bang and conclude this theory is not as strong as people think. Indeed, the cosmos is nowhere near its conventional age. The Bible says the Lord Jesus made all creation from nothing several thousand years ago, not billions. Truly, the heavens are filled with evidence that the universe is young and declares the glory of God (Psalm 19:1).

God created the heavens and the earth (Jeremiah 10:12), and you're His masterpiece within it.

1
CREATION *EX NIHILO* THROUGH JESUS CHRIST

Jeffrey P. Tomkins, Ph.D.

The first verse of the Bible provides the foundational truth of our faith and the fundamental truth concerning the universe: "In the beginning God created the heavens and the earth" (Genesis 1:1). The phrase "heavens and the earth" means the entire universe.

In the Hebrews 11 list of what we are to understand by faith, the first thing noted is the creation of all things by the word of God: "By faith we understand that the worlds were framed by the word of God, so that the things which are seen were not made of things which are visible" (v. 3). Thus, we are to receive by faith the supreme truth that everything in this present universe was brought into existence literally out of nothing by the good pleasure and will of God.

One of the great articles of Christian faith, penned by 121 Puritan clergymen in 1649, succinctly states this foundational truth of Genesis 1.

It pleased God the Father, Son, and Holy

> Ghost, for the manifestation of the glory of his eternal power, wisdom, and goodness, *in the beginning, to create, or make of nothing, the world, and all things therein, whether visible or invisible,* in the space of six days, and all very good.[1]

This truth that God created the universe out of nothing is sometimes described by the Latin phrase *ex nihilo* (out of nothing). This ability of God to create *ex nihilo* is not only mentioned in Hebrews 11:3 but is also inferred in Romans 4:17, where God "calls those things which do not exist as though they did." The English Standard Version says He "calls into existence the things that do not exist." In Psalm 33, the Scriptures say, "By the word of the LORD the heavens were made, and all the host of them by the breath of His mouth....For He spoke, and it was done; He commanded, and it stood fast" (vv. 6, 9).

When God speaks something into existence that didn't previously exist, such as the universe, what is implied? Nothingness cannot beget nothingness. Truly, God is the great first cause and source of the entire universe. This is traditionally called the cosmological argument in apologetics.

What Is Creation *Ex Nihilo*?

What exactly is the fundamental meaning of the biblical doctrine of *ex nihilo*? This Genesis-based proposition answers two age-old questions: "What is the material cause of the universe?" and "What did God make the world out of?" In short, the answer to both questions is "nothing." But what is "nothing"?

Philosophically speaking, any definition or paradigm describing nothing could actually make it something.

Some philosophers and theologians have thought that it could be more descriptive to speak of creation into nothing instead of out of nothing. While both ideas are helpful, we need to understand that creation is neither made "out of" a preexisting material nor created "into" a preexisting place. In this regard, theologian John Frame said, "We must oppose both the Aristotelian notion of an eternal 'matter' and the Platonic notion of an eternal 'receptacle.'"[2] These were popular ancient Greek philosophies.

Yet another false view is the proposition that the universe is merely an extension or emanation of God Himself. This idea is known as pantheism, where creation and God are believed to be one—making creation itself divine.

So, how can we best understand the idea that creation *ex nihilo* came from a transcendent God? Instead of saying that God created the world from nothing, it might be better to say that God created the universe without any preexisting material. As the Scriptures indicate, God spoke and things appeared, including space and time (a medium) for all created things to occupy and in which to function.

When we construct creation *ex nihilo* in this way, it requires no definition of "nothing"; it simply refutes the nonbiblical view that God created the universe from preexisting stuff. In this sense, creation *ex nihilo* can be thought of philosophically as a negative proposition that affirms that the world was neither made from preexisting material nor is a pantheistic extension of God's being. Thus, since there are no other alternatives, the universe was *not made from anything*—it was made from nothing (*ex nihilo*).

Evolution *Ex Nihilo*?

Amazingly, evolutionists have often criticized creationists for the doctrine of creation *ex nihilo* being unscientific while they maintain their own version of evolution *ex nihilo* without a creator or logical first cause.[3] What hypocrisy! The biblical creation doctrine quite logically maintains that an infinite, omnipotent, omniscient, transcendent, self-existing God who upholds and personally interacts with His creation (is immanent) is the source and first cause of the universe. Evolutionists, however, have no reasonable source or first cause for a paradigm of Big Bang *ex nihilo* cosmology (evolution *ex nihilo*). Even worse, a considerable amount of astronomical scientific data has now refuted the so-called evidence for the Big Bang and an ancient universe.[4–10]

Edward P. Tryon was one of the first modern physicists to promote evolution *ex nihilo*. He recounted the whole affair, saying:

> In 1973, I proposed that our Universe had been created spontaneously from nothing (*ex nihilo*), as a result of established principles of physics. This proposal variously struck people as preposterous, enchanting, or both.[11]

In a 2002 *Discover* magazine article, the authors well-summarized the idea of evolution *ex nihilo* by highlighting an interview with theoretical physicist

Alan Guth.

> The primordial "stuff" of inflation [aftereffect of the Big Bang], he and other cosmologists contend, is very likely a spontaneous creation, a no-strings gift that boiled out of absolutely nowhere by means of an utterly random but nonetheless scientifically possible process.[12]

Theoretical physicist and cosmologist Lawrence Krauss published a book in 2012 titled *A Universe from Nothing*. The final chapter was contributed by well-known atheist Richard Dawkins.[13] The ideas Krauss promoted in his book and other sources have received strong criticism, with astronomer Luke Barnes saying:

> First and foremost, I'm getting really rather sick of cosmologists talking about universes being created out of *nothing*. Krauss repeatedly talked about universes coming out of nothing, particles coming out of nothing, different types of nothing, nothing being unstable. This is nonsense.[14]

Quantum Fluctuations and Nothingness

What Krauss and other theoretical physicists are really talking about when they refer to nothingness is the theoretical idea of a quantum vacuum. If a quantum vacuum did actually exist (according to theoretical physicists), it would not be nothing but a type of something. It would have properties, it would have energy (albeit very little), it would fluctuate, it could contribute to the alleged expansion of the universe (now in doubt), and it would obey the complex equations of quantum field theory. In other words, the quantum vacuum is not "nothing," and this type of hypothesis leads to a view that considers the sudden appearance of the universe as essentially nothing more than a fortuitous accident.

Nevertheless, theoretical physicists have seemed to be quite enamored with the beginning of the universe "*ex nihilo*" from a quantum vacuum, starting, as mentioned, in 1973 with Edward Tyron, who said, "So I conjectured that our Universe had its physical origin as a quantum fluctuation of some pre-existing true vacuum, or state of nothingness."[11] Even more absurd is a statement made by Alan Guth that "theoretically, anything—a dog, a house, a planet—can pop into existence by means of this quantum quirk, which physicists call a vacuum fluctuation."[12] Krauss said that "we all literally emerged from quantum nothingness," and "this is so strikingly nonintuitive that it can seem almost magical."[13]

Henry M. Morris, the founder of the Institute for Creation Research, best summarized the efforts of modern evolutionary cosmologists when he said:

> Regardless of the sophisticated mathematical apparatus leading the inflationary-universe cosmogonists to their remarkable statement of faith in the omnipotence of nothingness, there will continue to be a few realists who prefer the creationist alternative: "In the beginning God created the heaven and the earth."[3]

A Better Christ-Based Alternative

In addition to the opening chapters of Genesis, we find more information regarding creation in the New Testament. At the beginning of John's gospel, we're told concerning Jesus Christ, the God-man, that "all things were made through Him, and without Him nothing was made that was made" (John 1:3). The first clause in this verse in the Greek text literally reads "all things through Him came to be," with the Greek word for "all things" (*panta*) referring to the entire universe.

The apostle Paul is even more explicit in Colossians when he specifies that all the things in the universe, visible or invisible, were made by and through Christ Jesus, saying, "For by Him all things were created that are in heaven and that are on earth, visible and invisible....All things were created through Him and for Him" (Colossians 1:16).

Creation and Worship

God is our sovereign Lord and Creator, and He rules over all the universe. There is nothing in creation that's to be worshiped instead of or in addition to Him (e.g., nature). However, if we were to deny creation *ex nihilo*, we would logically have to affirm that some sort of matter has always existed and that it is eternal like God. This erroneous idea is an affront to God's independence and sovereignty and the fact that our worship is due to Him only. And if matter or living creatures existed apart from God, then what inherent right would God have to rule over creation and govern it for His glory?

But God did create the universe *ex nihilo*—out of nothing. Therefore, the universe and our lives have meaning and purpose, and God has revealed that purpose to us in His holy Word, the Bible, and in His coming as God in the flesh—the Lord Jesus Christ. Let us serve and worship this mighty Creator and meditate upon His Word.

References

1. Westminster Assembly. *The Westminster Confession of Faith: Edinburgh Edition*. 1649, reprinted 1985. Free Presbyterian Publications, 31. Emphasis added.
2. Frame, J. M. 2013. *Systematic Theology: An Introduction to Christian Belief*. Phillipsburg, NJ: P&R Publishing, 193.
3. Morris, H. M. 2011. Evolution *Ex Nihilo*. *Acts & Facts*. 40 (9): 4–5.
4. Hebert, J. and B. Thomas. 2014. Does Science Support the Big Bang? *Acts & Facts*. 43 (7): 21.
5. Hebert, J. 2018. Does the Cosmic Microwave Background Confirm the Big Bang? *Acts & Facts*. 47 (6): 10–12.
6. Hebert, J. 2018. Our Young Solar System. *Acts & Facts*. 47 (9): 10–13.
7. Hebert, J. 2019. Deep-Space Objects Are Young. *Acts & Facts*. 48 (9): 10–13.
8. Hebert, J. 2019. Continuing Troubles for the Big Bang Model. *Acts & Facts*. 48 (11): 10–13.
9. Hebert, J. 2021. Does the Universe Look Old? *Acts & Facts*. 50 (10): 19.
10. Hebert, J. 2022. James Webb Telescope vs. the Big Bang. *Acts & Facts*. 51 (7): 14–17.
11. Tryon, E. P. 1984. What Made the World? *New Scientist*. 101: 14.
12. Lemley, B. and L. Fink. 2002. Guth's Grand Guess. *Discover*. 23 (4): 32–39.
13. Krauss, L. M. 2012. *A Universe from Nothing*. New York, NY: Free Press.
14. Barnes, L. Of Nothing. Letters to Nature. Published on letterstonature.wordpress.com April 1, 2011, accessed October 27, 2022. Emphasis in original.

2
A SUPERNOVA AND THE SCRIPTURE

Brian Thomas, Ph.D.

Well, there goes another star, disappearing into the night as if it had never existed. For an entire year, Hubble scientists used the space telescope to record snapshots of SN 2018gv—a supernova (SN) or exploded star. Experts then edited those images into a NASA video showing the fast fading of the supernova.[1] It's hard not to wonder, given the astronomical number of stars in the night sky, why we see more stars fade than form. For that matter, has anyone *ever* seen a star form?

Koichi Itagaki, an amateur astronomer, first found the exploding star SN 2018gv as a super bright spot amid spiral galaxy NGC 2525 in January 2018.[1] Professional astronomers then trained the Hubble Space Telescope onto the fast-fading object. Reporting on the video, NASA news described the fading process:

> The type of supernova seen in this sequence originated from a burned-out star—a white dwarf located in a close binary system—that is accreting material from its companion star. When the white dwarf reaches a critical mass,

> its core becomes hot enough to ignite nuclear fusion, turning it into a giant atomic bomb. This thermonuclear runaway process tears the dwarf apart. The opulence is short-lived as the fireball fades away.[1]

At a basic level, therefore, we see stars fade. But do we see stars *form*? Conventional scientists believe stars still form today, but they have never directly observed it. Misleading claims of "newborn" stars refer merely to stars that are *thought* to have formed more recently than most—no one has actually witnessed it.[2] No astronomer—amateur or professional—has yet reported detecting a spot in space where a star ignited. News outlets would doubtless trumpet such an event. So, if we don't see them form naturally, then how did they get there?

With so many people owning telescopes these days, shouldn't some folks have recorded stars forming by now? Conventional scientists estimate that hundreds of new stars should be forming just within our own Milky Way Galaxy every year. But the few they tag as freshly formed stars were already shining.[3] A steady string of new lights blinking on would confirm the concept that our universe's many stars formed with no help from any Creator. Instead, we only see them blinking out.

Thus, stars fade but they don't form, and the Bible indicates the same. Genesis 1:16 reports star formation in the past tense, as though God made stars through a miracle. Based on this, we should not expect to see star formation, and we don't.[4]

G299, a supernova remnant in the Milky Way galaxy

Nor does the Bible speak of stars forming at later points. It does, however, speak of a cursed creation, where features fade and creatures crumble. Romans 8:22 says, "For we know that the whole creation groans and labors with birth pangs together until now." Those birth pangs, including fast-fading supernovas, point to God's promise to create a whole new heavens and earth at some future moment and do not refer to a continuous birth of new stars today. Biblically, we should expect to see stars fading, and we do. While the Bible is not an astronomy textbook, the way that Scripture deals with the history of stars is confirmed by modern telescopes.

Everything the Bible touches on, including the human heart, also fits what we observe. Jeremiah 17:9 describes the heart as "deceitful above all *things*, and desperately wicked; who can know it?" With stars blinking out and not on, we're going to need new stars that don't fade but will burn forever. And with hearts full of sin that earns us death, we're going to need new and sinless hearts that last forever.[5] The Bible promises both to those who trust in Jesus.

SN 2018gv reminds us that the Bible got it right when it comes to an originally good creation that God cursed because man sinned. This same Bible says that heavenly luminaries will fall away before God creates the new heavens. [6,7] It will be a fitting new universe into which "penitents with righteousness" can carry their new hearts.[8]

References

1. Garner, R., ed. Hubble Watches Exploding Star Fade Into Oblivion. Posted

on NASA.gov October 1, 2020, accessed October 7, 2020.

2. Mathewson, S. Astronomers spot hundreds of baby stars and planet-forming disks. *Space.com*. Posted on space.com March 1, 2020, accessed October 8, 2020.

3. Cline, J. and G. Novak. How a Star is Born (And Why We Don't Know Why). *Helix*. Posted on helix.northwestern.edu August 14, 2009, accessed October 8, 2020.

4. Even Big Bang proponent and astronomer Neil deGrasse Tyson has acknowledged that "if none of us knew in advance that stars exist, front line research would offer plenty of convincing reasons why stars could never form." See Tyson, N. D. 2007. *Death by Black Hole: And Other Cosmic Quandaries*. New York: W. W. Norton & Company, Inc., 187.

5. Romans 6:23.

6. Isaiah 34:4.

7. Isaiah 65:17.

8. Isaiah 1:27.

First-ever photo of a black hole at the center of the Milky Way galaxy

3
FIRST-EVER PHOTO OF A BLACK HOLE

Jake Hebert, Ph.D.

On April 10, 2019, a team of astronomers revealed the first-ever photograph of a black hole.[1] Modern physics sees time and space as interconnected, forming a kind of "fabric" or "membrane" called spacetime. A black hole is a region of spacetime that is extremely distorted by the presence of a sufficiently dense mass. The gravity in this region is so strong that anything getting too close to the black hole will never escape.

Some black holes have masses comparable to that of our sun. These "stellar mass" black holes are thought to form when a star consumes the last of its nuclear fuel and collapses under the influence of its own gravity. Other supermassive black holes are thought to have masses millions or even billions of times greater than this. These supermassive black holes seem to be present at the centers of most, if not all, large galaxies, including our own Milky Way.[2]

This particular black hole is a supergiant black hole that has an estimated mass five billion times greater than that of our sun. It's located in the "supergiant" elliptical galaxy M87 about fifty million light-years from Earth.[1]

Scientists have long speculated about the possible existence of such objects, and there is much evidence for their existence. However, this is the first time that a black hole has been imaged directly.

...Or at least somewhat directly. The black hole itself can't really be photographed because its gravity is so strong that any light emitted from within the black hole's edge (a spherical boundary called its event horizon[3]) literally cannot escape. Rather, the image shows a glowing disk of gas surrounding the black hole's shadow. This swirling disk of gas has been "swept up" by the black hole's strong gravity. Scientists performed this very impressive technical achievement by using a network of eight radio telescopes around the world called the Event Horizon Telescope.[1]

Does this have any relevance for creation? Due in part to the weird nature of black holes, some creation supporters have questioned whether they are real. We have every reason to think that they are. Einstein's theory of relativity makes very accurate predictions about their behavior and, as stated earlier, there is much indirect—and now direct—evidence for their existence. The image agrees very well with predictions from Einstein's theory of general relativity.[4]

Finally, it's worth noting that supermassive black holes, such as the one just photographed, present big problems for Big Bang scientists. Some of these giant black holes have been detected at enormous distances from us. In Big Bang thinking, this means we are seeing these giant objects, not as they are now, but as they were shortly after the Big Bang. The most extreme example is a giant black hole thought to have formed just 690 million years after the Big Bang.[5] Yet conventional scientists are puzzled as to how natural processes could form such massive objects so relatively quickly.[6,7] This is very similar to the problem of mature distant galaxies that also plagues the Big Bang model.[8]

Black holes may be weird, but like all God's works, they defy conventional attempts to explain the universe apart from our Creator.

References

1. Grossman, L. and E. Conover. The first picture of a black hole opens a new era of astrophysics. *Science News*. Posted on sciencenews.org April 10, 2019, accessed April 10, 2019.
2. Active Galactic Nuclei (AGN)/Supermassive Black Holes. Website of the Spitzer Space Telescope at the Jet Propulsion Laboratory, California Institute of Technology. Accessed April 10, 2019.
3. The term "event horizon" was coined by Dr. Wolfgang Rindler, one of the professors at the University of Texas at Dallas. Kotaru, A. UTD Celebrates Legacy of Founding Faculty Member. *The Mercury*. Posted on utdmercury.com February 25, 2019, accessed April 10, 2019.
4. As an aside, many creation researchers think that Einstein's theory can help explain how we can see distant starlight in a young universe.
5. Landau, E. and E. Bañados. Found: Most Distant Black Hole. Jet Propulsion Laboratory. Posted on jpl.nasa.gov December 6, 2017, accessed April 10, 2019.
6. Kormendy, J. and L. C. Ho. 2013. Coevolution (Or Not) of Supermassive Black Holes and Host Galaxies. *Annual Review of Astronomy and Astrophysics*. 51: 511–653.
7. Saplakoglu, Y. Zeroing In on How Supermassive Black Holes Formed. *Scientific American*. Posted on scientificamerican.com September 29, 2017, accessed April 10, 2019.
8. Thomas, B. Distant Galaxies Look Too Mature for Big Bang. *Creation Science Update*. Posted on ICR.org November 30, 2011, accessed April 10, 2019.

4
CONTINUING TROUBLES FOR THE BIG BANG MODEL

Jake Hebert, Ph.D.

Cosmology is the study of the origin and structure of the universe, and the Big Bang is the dominant conventional cosmological model. Some Christians say God used the Big Bang to create the universe, but that model contradicts Scripture on multiple points.[1] There have been some recent developments involving the Big Bang model, nearly all of which are bad news for its proponents.

According to the Big Bang model, the universe was once very dense and hot. Supposedly, the universe began expanding rapidly about 14 billion years ago and is still expanding today. This expansion, inferred from clues within light from distant galaxies, is one of three main arguments for the model.[2] A second argument is that the Big Bang does a good job of accounting for the light chemical elements hydrogen and helium. A third is the existence of faint cosmic microwave background (CMB) radiation coming to us from all directions in space (Figure 1). Big Bang proponents interpret the CMB as an "afterglow"

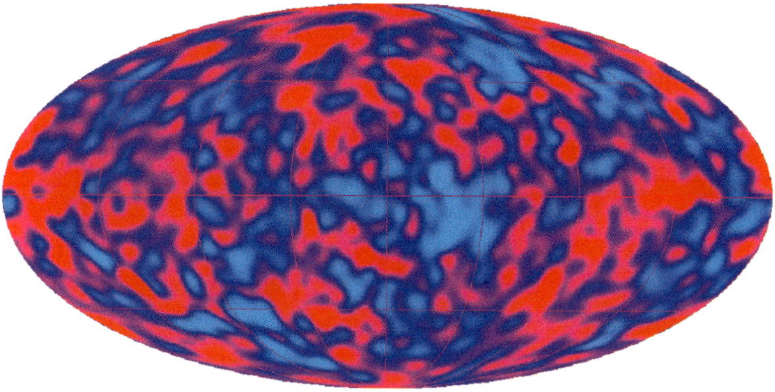

Figure 1. *Cosmic Background Explorer (COBE) satellite map of CMB temperature*

from a time about 400,000 years after the Big Bang occurred.

Despite these apparent successes, the Big Bang model has serious scientific problems. One enormous difficulty is that Big Bang proponents have concluded that about 95% of the "stuff" in the universe is composed of mysterious entities called dark matter and dark energy, but they don't know what these things are. How can Big Bang theorists claim to understand the process that supposedly brought the universe into existence when, by their own admission, 95% of the universe's contents are unknown?[3]

As a creation ministry, the Institute for Creation Research wants people to be up to date on the current version of the Big Bang model instead of one that was popular decades ago. For instance, Big Bang cosmologists used to say the universe went through an enormous "growth spurt" called inflation shortly after the Big Bang. However, most theorists today claim that inflation happened first and *caused* the Big Bang.[4]

Hubble Constant Contradiction Persists

Most astronomers think the universe is expanding, causing galaxies to move away from each other. Scientists use a number called the Hubble constant, denoted by the symbol H_0, to characterize this expansion. They use two different methods to calculate H_0. One way is to calculate the value directly, using estimated distances and speeds of distant galaxies. A second way is to infer this number by looking at details of the CMB radiation. The values calculated from these two methods conflict with each other, and a more recent study hasn't resolved the issue.[5–7]

When Big Bang proponents use the CMB to infer a value for H_0, they assume the Big Bang model is correct. Naturally, if the model is wrong, there's no reason to expect this method to yield an accurate result. Creationists aren't surprised these two different methods yield contradictory results. And even though the CMB is arguably the strongest argument for the Big Bang, there are details about this radiation that do *not* align with the Big Bang model.[8] For instance, Cambridge astrophysicist George Efstathiou commented on how the CMB doesn't match the expectations of inflation theory:

> The theory of inflation predicts that today's universe should appear uniform at the largest scales in all directions....That uniformity should also characterize the distribution of [temperature] fluctuations at the largest scales within the CMB. But these anomalies,

> which [the] Planck [satellite] confirmed, such as the cold spot, suggest that this isn't the case….This is very strange….And I think that if there really is anything to this, you have to question how that fits in with inflation….It's really puzzling.[9]

Missing Baryonic Matter Found?

Heavy subatomic particles like protons and neutrons are called baryons. Because protons and neutrons comprise nearly all the mass of an atom, the normal atomic matter we interact with in our everyday experiences is called baryonic matter.

As mentioned earlier, one of the three main arguments for the Big Bang is that it can account for the observed abundance of hydrogen and helium in the universe. However, this is because the model has an adjustable parameter, like a tuning dial on a radio.[10] Big Bang scientists choose a value for this parameter to ensure that the model matches the observed abundance of hydrogen and helium.[11]

So, contrary to popular perception, the Big Bang does *not* successfully predict the abundance of hydrogen and helium. Rather, the model's proponents *choose* a value for this parameter to make sure the model gives the right answer.[12–14] Nevertheless, conventional scientists consider the model's ability to match the observed abundance of hydrogen and helium to be a major success.

Once Big Bang scientists choose their value for this parameter, the model indicates how much bary-

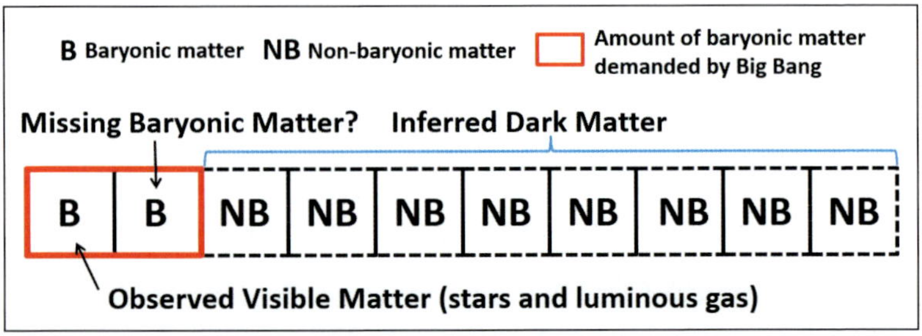

Figure 2. *After the Big Bang is "tuned" to give the correct abundance of hydrogen and helium, it produces enough baryonic matter to account for about 20% of the matter most astronomers think exists. The remaining 80% is assumed to be an exotic form of invisible "dark matter."*

onic matter should exist in the universe.[15] When one adds up the different forms of matter thought to exist, the amount of baryonic matter predicted by the Big Bang is only 20% of the total (Figure 2). Big Bang astronomers think the other 80% is an exotic form of invisible dark matter, discussed in the next section. Previous observations indicated that visible stars and gas could only account for half this predicted baryonic matter, and scientists couldn't account for the other half.

Astronomers claimed to have solved this problem in 2018.[16] (Interestingly, another scientist claimed to have solved it in 2017.[17]) Theorists think the missing baryonic matter should reside in thin, hot strings of ionized hydrogen located between galaxies. Astronomers didn't detect the hydrogen *per se* but rather ionized oxygen that they think is associated with the hydrogen. Naturally, Big Bang proponents will see this as good news for their model. However, it's important to realize that the missing matter hasn't actually been found directly. Rather, oxygen was found that conventional scientists think, based on their models, should be associated with the missing hydrogen.

It's worth noting that the Wikipedia entry for "Missing baryon problem" has been flagged for possibly making too strong a claim about the problem being solved, despite the obvious anti-creation bias found in Wikipedia articles touching on the creation-evolution controversy.[18]

Dark Matter Still Undetected

As mentioned in the last section, many astronomers think 80% of all the matter in the universe is invisible dark matter. Although astronomers deduced the existence of dark matter apart from the Big Bang model, this substance has become very important to conventional cosmologists. They recognize the enormous problems in their theories of star and galaxy formation. Many claim dark matter is the missing ingredient that can somehow enable their theories to

work.[19] This is very convenient for theorists. Since no one knows what dark matter is—or even if it really exists—no one can demonstrate that their theories are wrong.[20]

Because the Big Bang model only allows for 20% of all matter to be baryonic (made of atoms), its proponents must assume that dark matter is something else. Other forms of matter (e.g., free electrons, neutrinos, etc.) do exist but have generally been ruled out as dark matter candidates. The scientists have no choice but to insist that dark matter is some exotic, never-before-observed substance.

So, how is the hunt for this exotic matter going? Not well. Repeated searches have come up empty,[21] and theorists are becoming increasingly nervous if not desperate.

Dark Matter Before the Big Bang?

How desperate? One theorist recently suggested that perhaps dark matter somehow existed *before* the Big Bang.[22,23] How is that possible? Haven't we been led to believe that the Big Bang was the origin of *everything*?

This theorist said dark matter came from something called a scalar field that supposedly was present before the Big Bang. A problem with this idea is that only one scalar field is known to exist, and that's the field associated with the famous Higgs boson. All other scalar fields are hypothetical.

By the way, this should give pause to Christians who say God used the Big Bang to create the universe. If the supposed "bang" was God's initial creative act, then according to this reasoning, dark matter existed *before* Genesis 1:1. If 80% of all existing matter had an existence before then, did God actually create it prior to Genesis 1:1? If so, why doesn't the Bible tell us? If not, is dark matter simply eternal? And if it's eternal, what does that do to Christian theology?

Time Before the Big Bang?

This raises another point. Big Bang scientists had long insisted that speaking of time before the Big Bang was as nonsensical as asking the question "What is north of the North Pole?" Well, apparently the question wasn't as nonsensical as we were led to believe, because they now *routinely* talk about time before the Big Bang. In fact, inflation theorists now claim the inflation process that supposedly triggered the Big Bang could have been going on for eons by the time the Big Bang supposedly created our universe. This has led to the idea that our universe is only one of an infinite number of universes in a supposed multiverse.[24]

Conclusion

This should demonstrate just how "squishy" Big Bang theories are. Conventional scientists simply won't allow data to falsify them, even if it means tacking on additional hypotheses or accepting concepts that they themselves dismissed as nonsense decades ago, such as time before the Big Bang.

Instead of attempting to harmonize the inerrant Word of God with a flimsy scientific model, Christians would do far better to simply take God's Word at face value. The universe came into existence not through the Big Bang but by the omnipotent Word of God.

References

1. Morris, J. D. 1997. Is the Big Bang Biblical? *Acts & Facts.* 26 (5).
2. Some creationists (myself included) question whether this is a correct interpretation of the clues within this distant light. See Hartnett, J. 2011. Does observational evidence indicate the universe is expanding?—part 2: the case against expansion. *Journal of Creation.* 25 (3): 115–120.
3. Hebert, J. and B. Thomas. 2014. Does Science Support the Big Bang? *Acts & Facts.* 43 (7): 21.
4. Hebert, J. 2013. The Ever-Changing Big Bang Story. *Acts & Facts.* 42 (1): 14.
5. Hebert, J. Big Bang Hubble Contradiction Confirmed. *Creation Science Update.* Posted on ICR.org May 16, 2019, accessed August 23, 2019.
6. Freedman, W. L. et al. 2019. The Carnegie-Chicago Hubble Program. VIII. An Independent Determination of the Hubble Constant Based on the Tip of the Red Giant Branch. *Astrophysical Journal.* 882 (1): 34.
7. New Hubble Constant Measurement Adds to Mystery of Universe's Expansion Rate. *HubbleSite.* Posted on hubblesite.org July 16, 2019, accessed August 23, 2019.
8. Hebert, J. 2018. Does the Cosmic Microwave Background Confirm the Big Bang? *Acts & Facts.* 47 (6): 10–12.
9. Discoveries from Planck may mean rethinking how the universe began. The Kavli Foundation. Posted on phys.org July 26, 2013, accessed September 6, 2019.
10. This adjustable parameter is called the baryon-to-photon ratio. It is a measure of how many baryons are in the universe compared to the number of light "particles" (photons).
11. Bergström, L. and A. Goobar. 2008. *Cosmology and Particle Astrophysics,* 2nd ed. Chichester, UK: Springer Praxis Publishing, 167–176.
12. Krauss, L. M. 2012. *A Universe from Nothing: Why There Is Something Rather Than Nothing.* New York: Free Press, 24–25.
13. Hoyle, F., G. Burbidge, and J. V. Narlikar. 2000. *A Different Approach to Cosmology: From a Static Universe through the Big Bang towards Reality.* Cambridge, UK: Cambridge University Press, 97.
14. Even with this freedom, the Big Bang still cannot adequately account for the relative abundance of different isotopes of lithium. See Thomas, B. Big Bang Fizzles under Lithium Test. *Creation Science Update.* Posted on ICR.org September 22, 2014, accessed August 20, 2019.

15. Hebert, J. 2013. Dark Matter, Sparticles, and the Big Bang. *Acts & Facts*. 42 (9): 17–19.

16. Nicastro, F. et al. 2018. Observations of the missing baryons in the warm–hot intergalactic medium. *Nature*. 558: 406–409.

17. Crane, L. Half the universe's missing matter has just been finally found. *New Scientist.* Posted on newscientist.com October 9, 2017, accessed September 9, 2019.

18. Missing baryon problem. Wikipedia. Posted on wikipedia.com, accessed September 9, 2019.

19. Mo, H., F. van den Bosch, and S. White. 2010. *Galaxy Formation and Evolution*. Cambridge, UK: Cambridge University Press, 8, 15.

20. Psarris, S. 2012. *What You Aren't Being Told About Astronomy, Volume 2: Our Created Stars and Galaxies.* DVD. Creation Astronomy Media.

21. Letzter, R. Physicists Keep Trying—and Failing—to Find Dark Matter in Dark Places. *Live Science*. Posted on livescience.com December 7, 2018, accessed September 3, 2019.

22. Tenkanen, T. 2019. Dark Matter from Scalar Field Fluctuations. *Physical Review Letters*. 123, 061302.

23. Tantibanchachai, C. Dark matter may be older than the big bang, study suggests. John Hopkins University press release. Posted on hub.jhu.edu August 8, 2019, accessed September 3, 2019.

24. Hebert, J. 2013. Hiding from God in the Multiverse. *Acts & Facts.* 42 (6): 9.

Hubble Space Telescope

5
BIG BANG HUBBLE CONTRADICTION CONFIRMED

Jake Hebert, Ph.D.

Using data obtained from the Hubble Space Telescope, astronomers have confirmed a contradiction between two different estimates of the Hubble constant—an extremely important number in cosmology.[1]

Conventional scientists claim the universe is expanding as a result of a "big bang" 13–14 billion years ago. The inference of an expanding universe is based on the fact that the light spectra from distant galaxies have been "redshifted" to longer wavelengths. Some creation scientists think this interpretation of the redshift data is correct, while others do not. Even if the universe is expanding, this doesn't necessarily demand a Big Bang explanation. One could imagine scenarios in which God, for His own good purposes, imposed an expansion on His created universe, e.g., to prevent gravitational collapse. Of course, if redshifts are indeed caused by something other than cosmological expansion, creation scientists will need to provide an alternative explanation for those redshifts. In fact, creation scientists have already proposed alternative explanations, although they are still tentative at this point.[2]

The expansion rate is characterized by a figure called the Hubble parameter, also known as the Hubble constant, indicated by the symbol H_0. This number has units of speed per units of distance, namely kilometers per second per megaparsec—km/sec per mpc (a megaparsec is 3.26 million light-years). This means that the speed at which a galaxy is thought to be moving away from us depends on the galaxy's distance from us. The farther away the galaxy, the faster it's receding.[3]

Astronomers use different methods to infer the Hubble constant. The direct method uses distances and inferred speeds to calculate the Hubble constant. Previous studies used supernovae data, but a more recent study used 70 stars of varying brightness, called Cepheid Variables, within the Large Magellanic Cloud, a satellite galaxy of our own Milky Way galaxy. This method yields an estimate of 73–74 km/sec per mpc.[1,4]

In the indirect method, conventional scientists analyze details of the cosmic microwave background (CMB) radiation that they interpret as an "afterglow" from a time about 400,000 years after the Big Bang. They do so by constructing a theoretical Big Bang model containing a number of adjustable parameters whose values are left unspecified. Theorists then find the values of these parameters that give the best "overall" fit to the CMB data. From this, they then calculate the Hubble constant. This indirect method indicates that H_0 should be about 67 km/sec per mpc.[4] Of course, this approach implicitly assumes the Big Bang model is correct.

The two different estimates disagree, and the uncertainties in the estimates aren't large enough to explain the discrepancy. Scientists have known about this contradiction for years,[4,5] but the new measurements make it even less likely that the discrepancy between the two calculations is a mistake. They had concluded there was just a one in 3,000 probability that this result is in error. These new measurements reduce this probability to just one in 100,000.[1,6]

So, even if the universe is expanding, the Big Bang interpretation of the cosmic microwave background radiation yields an estimate of H_0 that contradicts the estimate obtained from more direct means. The paper's lead author, Adam Reiss, said, "The Hubble tension between the early and late universe may be the most exciting development in cosmology in decades."[1]

Well, that's one way to put a positive spin on negative results! Why are these results exciting? Because Big Bang expectations have been contradicted yet again? As creation researcher David Coppedge astutely noted, in *most* professions, being wrong over and over again isn't usually seen as "exciting"—it's a reason to lose your job![7]

This response is instructive because it shows how conventional scientists almost never allow data to count against the Big Bang, only *for* it. If the results agree with Big Bang expectations (which they almost never do), then this is claimed as a victory for the Big Bang. If the results disagree with Big Bang expectations, this isn't acknowledged as evidence against the model but rather is chalked up as evidence for some undiscovered physics to be added to the model.

Problems with the Big Bang are legion.[8,9,10] Conventional cosmologists have wasted and continue to waste who knows how many millions (perhaps billions?) of taxpayer dollars in their attempt to prop up a failing cosmology. Some Christians argue that God somehow used the Big Bang to create the universe, but the details of the Big Bang contradict the details given in Genesis,[11] and the Big Bang is a poor scientific model to boot!

References

1. Garner, R. Mystery of the universe's expansion rate widens with new Hubble data. *Phys.org*. Posted on phys.org April 25, 2019, accessed April 29, 2019.
2. Hartnett, J. Tension, not extension in creation cosmology. Creation Ministries International. Posted on creation.com August 12, 2014, accessed April 29, 2019.
3. Galaxy recession is thought to occur, not because the galaxies are moving *through* space away from us, but because space *itself* is expanding.
4. Hebert, J. Big Bang Continues to Self-Destruct. *Creation Science Update*. Posted on ICR.org April 25, 2016, accessed April 29, 2019.
5. Hebert, J. Big Bang Hubble Contradiction. *Creation Science Update*. Posted on ICR.org February 5, 2018, accessed April 29, 2019.
6. Reiss, A. G. et al. Large Magellanic Cloud Cepheid Standards Provide a 1% Foundation for the Determination of the Hubble Constant and Stronger Evidence for Physics Beyond ΛCDM. *The Astrophysical Journal*. Revised version of pre-print posted on arxiv.org March 27, 2019, accessed April 29, 2019.
7. Coppedge, D. F. Don't Trust Scientists Who Were Wrong; Fire Them. *Creation Evolution Headlines*. Posted on crev.info June 25, 2017, accessed April 29, 2019.

8. Thomas, B. Big Bang Fizzles Under Lithium Test. *Creation Science Update*. Posted on ICR.org September 22, 2014, accessed April 29, 2019.

9. Hebert, J. Big Bang Blowup at Scientific American. *Creation Science Update*. Posted on ICR.org May 29, 2017, accessed April 29, 2019.

10. Hebert, J. A Cosmic 'Supervoid' vs. the Big Bang. *Creation Science Update*. Posted on ICR.org May 7, 2015, accessed April 29, 2019.

11. Morris, J. D. 1997. Is the Big Bang Biblical? *Acts & Facts*. 26 (5).

6
NEW YORK TIMES EDITORIAL: BIG BANG UNRAVELING?

Jake Hebert, Ph.D.

Two physicists suggested in a 2023 *New York Times* guest editorial that Big Bang cosmology "may be starting to unravel."[1] Dr. Adam Frank (University of Rochester) and Dr. Marcelo Gleiser (Dartmouth College) gave an overview of three major problems confronting mainstream cosmology—problems that are already well known to well-informed creationists.

In their editorial, Drs. Frank and Gleiser acknowledged that data from the James Webb Space Telescope show that the most distant galaxies seem much too mature to agree with Big Bang expectations. Big Bang proponents assume that light from the most distant galaxies took many billions of years to reach us. Although this may seem reasonable, creationists think this assumption is open to challenge.[2] Given that assumption, though, we are not seeing these very distant galaxies as they are now but as they were more than 13 billion years ago. Hence, these very distant galaxies should look very immature and "unevolved." Yet these expectations are routinely contradicted,[3] and data from the Webb telescope have greatly exacerbated the problem.[4,5] Drs. Frank and Gleiser said the problem is "akin to parents and their children appearing in a story when the grandparents are still children themselves."[1] The problem is so severe that one physicist suggested doubling the assumed age of the universe to try to fix it.[6]

They also pointed out that two different methods of estimating the Hubble constant, one of the most important numbers in cosmology, give contradictory values. One of the methods uses the Big Bang interpretation of the cosmic microwave background (CMB) radiation, so this discrepancy may imply that Big Bang cosmologists are misinterpreting the CMB data.[7]

They noted some of the philosophical baggage accompanying modern inflation theory, like the idea of a multiverse containing an infinite number of unobservable universes. Inflation was tacked on ad hoc to the Big Bang to solve serious problems, one of which was the Big Bang's own version of the "distant starlight problem."[8] Yet there is zero evidence for inflation, and it has created more problems for Big Bang cosmologists.[9]

Just one year ago, a popular science writer said it was a "falsehood" that data from the James Webb telescope had disproved the Big Bang.[10] One could perhaps quibble over the use of the word "disproved" since Big Bang cosmologists have been extremely adept over the years at absorbing challenges to their model. True Big Bang believers can always add still more parameters to their model ad hoc so that the Big Bang will *never* be disproven or falsified, regardless of any data accumulated. But is that really science? The same writer who suggested that Big Bang critics were being deceptive also said that the "Big Bang theory is still on solid ground."[10] The *New York Times* editorial shows that the Big Bang critics were and are correct: the model is *not* on solid ground, despite happy talk to the contrary.

Some argue that Christians should accept the Big Bang model because it implies the universe had a beginning—which is consistent with Genesis in a very minimal sort of way. But the Big Bang contradicts Scripture in almost every other way, and some Big Bang cosmologists, like the late Stephen Hawking, have been working to tweak the model to do away with a beginning for the universe. What then for Christians who have tied their apologetics to the Big Bang?

The Lord Jesus Christ, not an alleged Big Bang, deserves the credit and glory for the amazing universe in which we live. How much better to believe His testimony recorded in Genesis than to trust the speculative, ever-changing stories of those determined to deny our Creator!

References

1. Frank, A. and M. Gleiser. The Story of Our Universe May Be Starting to Unravel. *New York Times*. Posted on nytimes.com September 2, 2023, accessed September 9, 2023.
2. Humphreys, D. R. Toward a More Biblical Cosmology. Gateway Creation Conference. Posted on youtube.com November 22, 2021, accessed September 11, 2023.
3. Hebert, J. Most Distant Rotating Disk Galaxy Challenges Secular Models.

Creation Science Update. Posted on ICR.org May 28, accessed September 9, 2023.

4. Hebert, J. 2022. James Webb Telescope vs. the Big Bang. *Acts & Facts*. 51 (11): 14–17.
5. Hebert, J. Webb Telescope Continues to Challenge Big Bang. *Creation Science Update*. Posted on ICR.org January 26, 2023, accessed July 12, 2023.
6. Hebert, J. Big Bang Troubles? No Problem Just Double the Universe's Age! *Creation Science Update*. Posted on ICR.org July 31, 2023, accessed September 9, 2023.
7. Hebert, J. 2018. Does the Cosmic Microwave Background Confirm the Big Bang? *Acts & Facts*. 47 (6): 10–14.
8. Coppedge, D. F. 2007. The Light-Distance Problem. *Acts & Facts*. 36 (6).
9. Hebert, J. 2017. Big Bang Blowup at Scientific American. *Creation Science Update*. Posted on ICR.org May 29, 2017, accessed September 11, 2023.
10. Cooper, K. The James Webb Space Telescope never disproved the Big Bang. Here's how that falsehood spread. *Space.com*. Posted on space.com September 7, 2022, accessed September 11, 2023.

7
WEBB TELESCOPE CONTINUES TO CHALLENGE BIG BANG

Jake Hebert, Ph.D.

Data obtained by the James Webb Space Telescope (JWST) continue to challenge expectations of Big Bang proponents.[1,2]

The JWST is designed to "see" in the infrared portion of the electromagnetic spectrum, just outside the range of visible light. Since redshifts[3] "push" most of the light from distant galaxies into the infrared part of the spectrum, the JWST can obtain better images of these very distant galaxies than if it were viewing them in visible light. In fact, the images obtained by the JWST are even more spectacular than the brilliant images obtained by the Hubble Space Telescope.

As noted in a previous *Creation Science Update*[4] and *Acts & Facts* article,[5] data from the JWST show that distant galaxies look more mature and developed than predicted by Big Bang theorists. Big Bang astronomers assume light from the most distant galaxies took almost 14 billion years to reach Earth. Although this may sound reasonable, creation scientists think this assumption is open to challenge.[6] But in

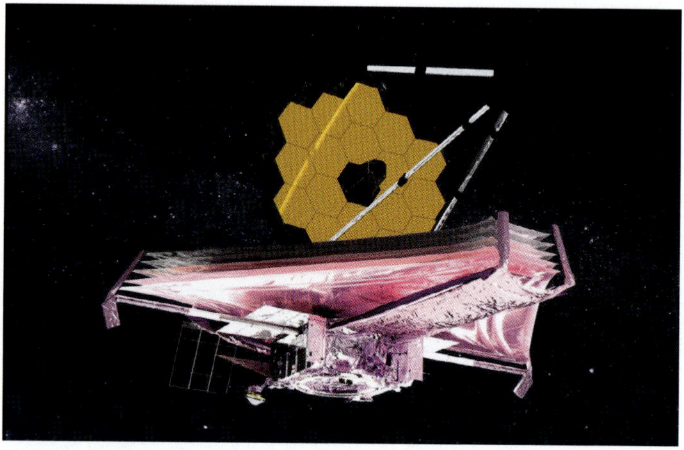

James Webb Space Telescope

any case, by Big Bang reckoning we should be seeing these very distant galaxies not as they are now but as they were almost 14 billion years ago. Hence, these galaxies should appear very "unevolved" and immature. Yet, this expectation is routinely contradicted,[7–9] and data from the JWST are making the disagreement even worse.

Big Bang theorists are being forced to quickly revise their predictions. In a 2022 article, I noted that in 2012 NASA was claiming that the universe's first stars formed about 400 million years after the Big Bang.[5] Actually, I was overly generous in my estimate: NASA was apparently making this claim as recent as 2021.[10] Yet, a *Nature* article published in 2022 stated that the first stars formed "perhaps around 250 million years after the Big Bang."[11] So within just the last two years, Big Bang theorists have had to push back the estimated "date" for the first stars by about 150 million years.

The supposed dates of the formation of the earliest galaxies are also being pushed back. That same *Nature* article stated, based on analysis of JWST data, that galaxies were already in existence "a few hundred million years [i.e., about 300 million years] after the Big Bang."[11] Yet Big Bang scientists then claimed, just months later, that the first galaxies could have formed as early as 200 million years after the Big Bang.[1]

Are you seeing a pattern? Big Bang theorists have to keep "pushing back" the times at which the first stars and galaxies supposedly appeared, because no matter how deep we look into space, stars and galaxies already exist! This puts increasing stress on the Big Bang model. Mainstream astronomers claim that stars and galaxies form naturally over millions of years and that the Big Bang occurred 13.8 billion years ago. So, by their reckoning, no star or galaxy can possibly be more than 13.8 billion years old. But as the supposed ages of the first stars edge closer and closer to 13.8 billion years, there is less and less time for natural processes to "make" stars and galaxies, even if one ignores the scientific problems with naturalistic star and galaxy formation theories.

Second, more powerful telescopes should allow astronomers to look deeper and deeper into space and (presumably, by Big Bang reckoning) further and further back in time. If the Big Bang is correct, telescopes may become powerful enough that we *should* be able to see "times" when no stars or galaxies exist. But what if stars and galaxies exist no matter how far out we look? What if no part of the visible universe fails to contain stars or galaxies? Will Big Bang proponents admit that the Big Bang has been disproven?

Despite the denials of a world that is in increasing rebellion against its Creator, the heavens do indeed testify of God's glory and their supernatural creation by the Lord Jesus Christ, exactly as described in Scripture.[12]

References

1. Stann, E. Astronomers suggest more galaxies were formed in the early universe than previously thought. *Phys.org*. Posted on phys.org January 9, 2023, accessed January 10, 2023.
2. Auburn, L. New study reveals wide diversity of galaxies in the early universe. *Phys.org*. Posted on phys.org January 9, 2023, accessed January 10, 2023.
3. A redshift is a stretching of wavelengths of light toward longer (or redder) wavelengths. This stretching is usually attributed to an inferred expansion of the universe.
4. Hebert, J. James Webb Telescope Data: Challenges for the Big Bang? *Creation Science Update*. Posted on ICR.org September 5, 2022, accessed January 10, 2023.
5. Hebert, J. 2022. James Webb Telescope vs. the Big Bang. *Acts & Facts* 51 (11): 14–17.
6. Humphreys, D. R. Toward a More Biblical Cosmology. Gateway Creation Conference. Posted on youtube.com November 22, 2021, accessed January 10, 2023.
7. Thomas, B. 'Old' Galaxy Found in 'Young' Part of the Universe. *Creation Science Update*. Posted on ICR.org May 24, 2011, accessed August 22, 2022.
8. Hebert, J. 'Early' Spiral Galaxy Surprise. *Creation Science Update*. Posted on ICR.org May 1, 2020, accessed August 22, 2022.
9. Hebert, J. Most Distant Rotating Disk Galaxy Challenges Secular Models. *Creation Science Update*. Posted on ICR.org May 28, 2020, accessed August 22, 2022.
10. A NASA Big Bang timeline image, first produced around 2006, states that the first stars formed about 400 million years after the Big Bang. However, the age of the universe of 13.77 billion years cited on the poster was calculated and published in 2021. Thus, NASA claimed that the universe's first stars formed about 400 million years ago.
11. Witze, A. Four revelations from the Webb telescope about distant galaxies. *Nature*. Posted on nature.com July 27, 2022, accessed January 10, 2023.
12. Genesis 1; Psalm 19:1–2; John 1:1–2; and Colossians 1:16.

8
ASTRONOMY MAGAZINE: BIG BANG IN CRISIS?

Jake Hebert, Ph.D.

The May 2020 issue of *Astronomy* magazine asks what might have once been seen as an unthinkable question: Is the Big Bang in crisis?[1] The article cites four major problems with the model:

1. the Big Bang implies that 95% of the universe's content is unknown to us,
2. the inability of the Big Bang to explain the enormous matter/antimatter imbalance in the universe,
3. the nature of cosmic inflation that was "tacked on" to the model years ago, and
4. contradicting values of the universe's inferred expansion rate.

There are three main arguments for the Big Bang model:

1. the apparent expansion of the universe,
2. the model accounts for the relative cosmic abundances of hydrogen and helium, and
3. the existence of the cosmic microwave background (CMB) that Big Bang theo-

rists interpret as an "afterglow" from a time about 400,000 years after the supposed Big Bang.

Not one of these arguments is an unqualified success, and apparent successes often lead to problems for the model.

Astronomers have inferred that other galaxies are receding away from our own, and they believe this is an indication that space itself is expanding. Big Bang theorists assume that this inferred expansion can be "rewound" backward in time to the supposed birth of our universe. Some creation scientists question whether astronomers are interpreting the redshift data correctly.[2] Others think the universe could indeed be expanding and that for some reason God imposed an expansion on His created universe. But in any case, an expanding universe is seen as one of the three main arguments for the Big Bang. Yet when Big Bang astronomers attempt to calculate the expansion rate using two different methods, they obtain contradictory results.[3]

Likewise, the ability of the Big Bang to account for the abundance of hydrogen and helium is not nearly as impressive as it first seems. The Big Bang model has an adjustable parameter,[4] like a tuning knob on a radio, that theoreticians can adjust to ensure that the Big Bang produces the amounts of hydrogen and helium necessary to match observations.[5,6] However, once the value for this parameter is chosen, the Big Bang model indicates how many atoms should exist in the universe. This is only 15–20% of the matter

Andromeda galaxy

that most astronomers think exists, so the remaining 80% or so has to be something else. Because Big Bang theorists have ruled out other known forms of matter, they are forced to hypothesize the existence of exotic, never-before-detected forms of non-baryonic matter. Needless to say, the search for this exotic dark matter is going very poorly.[7]

Moreover, Big Bang theorists acknowledge that, by their own reckoning, they do not know the composition of 95% of the mass/energy of the universe.[1] Or to put it another way, they don't know what the universe is composed of. This one simple fact demonstrates the absurdity of their claim that they know how the universe came to be. It is akin to claiming you understand the recipe for a cake, even though you don't know what kind of cake the recipe produces!

The cosmic microwave background radiation

(CMB) is arguably the strongest argument for the Big Bang, but there are features about the CMB that do not match Big Bang expectations nor the expectations of inflation theory, which has become a major part of the model. One Cambridge astrophysicist said,

> The theory of inflation predicts that today's universe should appear uniform at the largest scales in all directions....That uniformity should also characterize the distribution of fluctuations at the largest scales within the CMB. But these anomalies, which Planck confirmed, such as the cold spot, suggest that this isn't the case....[T]his is very strange.... And I think that if there really is anything to this, you have to question how that fits in with inflation....It's really puzzling.[8]

A 2004 open letter published in *New Scientist* had this to say about the Big Bang:

> Big bang theory relies on a growing number of hypothetical entities—things that we have never observed. Inflation, dark matter and dark energy are the most prominent. Without them, there would be fatal contradictions between the observations made by astronomers and the predictions of the big bang theory. In no other field of physics would this continual recourse to new hypothetical objects be accepted as a way of bridging the gap between theory and observation. It would, at the least, raise serious questions about the validity of the underlying theory.[9]

Creation scientists heartily agree. The scientific problems with the Big Bang model are so great that it should have been abandoned by theorists long ago. What is truly perplexing is why so many Christians insist that the Big Bang was God's means of creating the universe when it clearly contradicts the Genesis account in so many ways.[10]

References

1. Hooper, D. Is the Big Bang in Crisis? *Astronomy*. Posted on astronomy.com May 14, 2020, accessed May 15, 2020.
2. Hartnett, J. 2011. Does observational evidence indicate the universe is expanding?—part 2: the case against expansion. *Journal of Creation*. 25 (3): 115–120.
3. Hebert, J. Big Bang Hubble Contradiction Confirmed. *Creation Science Update*. Posted on ICR.org May 16, 2019, accessed May 18, 2020.
4. This adjustable parameter is called the baryon-to-photon ratio. It is a number that indicates the number of "heavy" particles (like protons and neutrons) in the universe compared to the number of quanta or "packets" of light.
5. Bergström, L. and A. Goobar. 2008. *Cosmology and Particle Astrophysics*, 2nd ed. Chichester, UK: Praxis Publishing, 167–176.
6. Hoyle, F., G. Burbidge, and J. V. Narlikar. 2000. *A Different Approach to Cosmology: From a Static Universe through the Big Bang towards Reality*. Cambridge, UK: Cambridge University Press, 97.
7. Hebert, J. Dark Matter Search Keeps Coming Up Empty. *Creation Science Update*. Posted on ICR.org April 9, 2020, accessed May 18, 2020.
8. Discoveries from Planck may mean rethinking how the universe began. The Kavli Foundation. Posted on phys.org July 26, 2013, accessed May 18, 2020.
9. Lerner, E. Bucking the Big Bang. *New Scientist*. Posted on newscientist.com May 22, 2004, accessed May 15, 2020. Also, see Open Letter on Cosmology.
10. Morris, J. D. 1997. Is the Big Bang Biblical? *Acts & Facts*. 26 (5).

NGC 3132, Southern Ring Nebula, viewed by NASA's Webb Telescope

9
JAMES WEBB TELESCOPE VS. THE BIG BANG

Jake Hebert, Ph.D.

The James Webb Space Telescope is thrilling astronomers with spectacular images of distant galaxies, but preliminary data contradicted Big Bang expectations. In order to understand these results, some background information is needed. I suggest, without being dogmatic about my personal position, a scientific test that creationists can perform that could clarify at least one basic issue in cosmology.

Galaxy Redshifts

Details in the light from distant galaxies are "shifted" to the red part of the electromagnetic spectrum (Figure 1). Astronomers use the letter z to denote the value of this redshift. Galaxy redshifts are a rough proxy or stand-in for distance—the higher the z value, the farther away the galaxy is

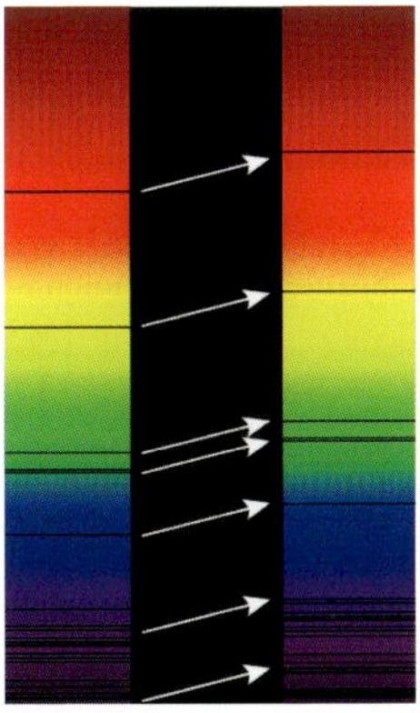

Figure 1. *Dark absorption lines from a cluster of galaxies are shifted toward the red part of the electromagnetic spectrum as compared to the same absorption lines in the sun's spectrum.*

from us. Galaxy redshifts are usually interpreted as evidence that space itself is expanding, causing galaxies to recede from us with speeds that increase with increasing z.

The James Webb Space Telescope (JWST) was constructed primarily to be sensitive to infrared radiation rather than visible light. Because light from very distant galaxies is redshifted to the infrared part of the spectrum, the JWST is obtaining better images of them than even the wonderful images provided by the Hubble Space Telescope (HST).

The Mature Distant Galaxy Problem

Big Bang theorists assume distant starlight takes billions of years to reach us. Although that may seem reasonable, creationists think this assumption is open to challenge.[1] But by Big Bang reckoning, it's implied we don't see very distant galaxies as they are now but as they were billions of years ago shortly after the Big Bang occurred. For this reason, theorists expect very distant galaxies to look extremely immature and "unevolved." However, this expectation is routinely contradicted.[2] Preliminary data from the JWST make this problem much more acute.

> For a long time...scientists believed the universe's earliest, oldest galaxies to be small, slightly chaotic, and misshapen systems. But...JWST-captured imagery has revealed these galaxies to be shockingly massive, not to mention balanced and well-formed—a finding that challenges, and will likely rewrite, long-held understandings about the

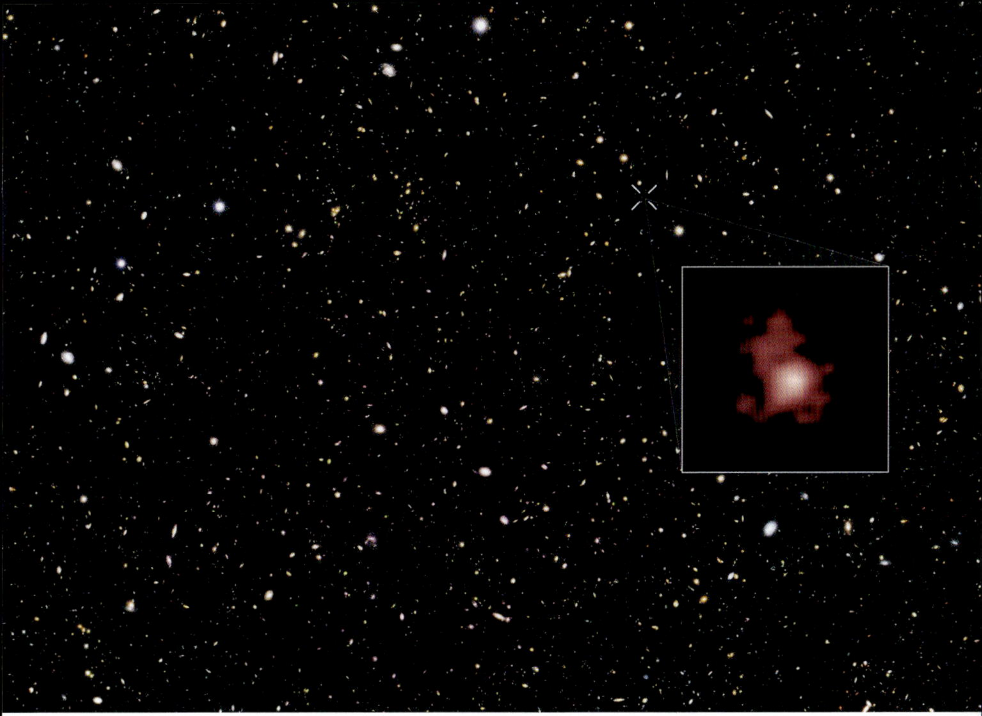

***Figure 2.** By evolutionary reckoning, the galaxy in the inset formed 400 million years after the Big Bang.*

origins of our universe.[3]

Evolutionary theorists now say the first stars formed between 100 and 250 million years after the Big Bang, and the first galaxies then formed a few hundred million years later.[4,5] However, about 10 years ago, they were saying it took 400 million years for the first stars to form.[6] So, a galaxy existing only 400 million years after the Big Bang puts a lot of stress on the theory. But the HST has already imaged such a distant galaxy, with a redshift of z = 11(Figure 2). With data from the JWST, astronomers are reporting galaxy redshifts as high as 16 and possibly even 20. These high redshift values imply that galaxies existed 250 million years (or less) after the Big Bang, in clear violation of the theory.[4]

Granted, these results are preliminary, but even these initial reports are causing angst among theorists. University College London cosmologist Richard Ellis said, "It worries me slightly that we find these monsters [galaxies] in the first few images."[7] University of California at Santa Cruz astronomer Garth Illingworth rhetorically asked, "How do you do this in the universe at such an early time? How do you form so many stars so quickly?"[3] And University of Kansas in Lawrence astronomer Allison Kirkpatrick said, "Right now I find myself lying awake at three in the morning wondering if everything I've ever done is wrong."[4]

The existence of very distant mature galaxies is just one more in a long list of failed Big Bang predictions.[8,9] Some popular science news sources now claim that pointing out these problems amounts to "science denial,"[10] but one only has to read the initial news reports to see that Big Bang scientists were greatly unsettled by these results.[11]

Unexpectedly Bright Distant Galaxies

This brings to mind another failed Big Bang expectation. Astrophysicist Michael J. Disney helped construct cameras for the HST. On his personal blog, he congratulated the James Webb team for their successful deployment of the JWST and recalled his experiences working on the HST.

> Before they design a single lens or circuit Space-camera teams must look deeply into the kinds of science they might be doing 20

> years ahead and my team asked me to look into high redshift galaxies. Knowing nothing of that subject at the time (1977) I naturally went round to ask the real experts. They were unanimous: "Don't bother with them because you'll never be able to see them owing to the Tolman effect."[12]

The Tolman effect is a surface brightness test that physicist Richard Tolman devised for an expanding universe. It involves the brightness of "extended" objects in space, objects that are not merely star-like points.[13] A galaxy's surface brightness is defined as its brightness divided by the area of angle measure that it "covers" in the sky. In a static, non-expanding universe, surface brightness should remain constant regardless of the value of (1+ z).[14] In an expanding universe, however, surface brightness should rapidly decrease with higher values of (1+ z). If galaxy surface brightness data fail to satisfy the Tolman test, this would imply that the redshifts, though real, are caused by something other than an expanding universe. Disney continued:

> Extended objects like galaxies would be dramatically dimmed by redshift in an expanding universe, so dimmed in practice as to become invisible in any decent Space camera. Imagine my shock then in 1993 when we got our first really deep Hubble image. It was plastered all over with very distant but quite distinct galaxies not dimmed in the least. That meant the universe couldn't be expanding—not according to Tolman.

> But when I tried to point this out the new self-appointed experts said; "No; it just means that young galaxies in the distant past were intrinsically much brighter than the galaxies today—after all they were younger in the past and generating more brilliant young stars."[12]

It's reasonable for Big Bang theorists to expect galaxy brightness to vary in a systematic way with increasing redshift since they believe galaxies evolve over time. So, in principle, theorists can claim these more ancient and distant galaxies were intrinsically brighter. However, Disney doesn't find this explanation convincing because he thinks it requires too much fine-tuning.[12] Nevertheless, he still apparently believes the universe is expanding and has proposed his own explanation for this unexpected result.[15]

The Tolman Test: Contradictory Results

Conventional astronomers, including the late Allan Sandage, claim galaxy surface brightness data are consistent with an expanding universe provided one assumes galaxy size and brightness evolve systematically with redshift.[16] However, Big Bang critic Eric Lerner claims that surface brightness data are much more consistent with a static universe and that it's impossible to reconcile the data with simple models of galaxy evolution when low-z nearby galaxies are included in the analysis.[17]

I don't presume to know who is correct. Sandage was a renowned, competent astronomer. On the other hand, Big Bang scientists were apparently surprised that the HST was able to see very distant galaxies. And one Big Bang astrophysicist has conceded that Lerner's results provide at least some support for the idea of a non-expanding universe.[18] But creationists must nail down this issue if we want to construct a biblical creationist cosmology. We cannot afford to take others' word for it, and we should perform our own Tolman test using all publicly available galaxy data. A negative result would indicate that redshifts are due to something other than an expanding universe. In that case, it would be very strong evidence against the Big Bang. But it would also mean that creation scientists have much work to do since we would need to provide a viable alternative explanation for galaxy redshifts, one that so far has remained elusive.

Redshifts in a Biblical Cosmology

But what if conventional astronomers are correct and the data *are* consistent with an expanding universe? Does this mean the Big Bang is in the clear? No. If redshifts do indeed represent the speeds of galaxies moving away from us, this can be interpreted to mean that our galaxy is at the center of the universe, a possibility acknowledged by even the late Stephen Hawking![19]

But Big Bang theory demands that there be no "special" places in the universe, including no unique center. Moreover, if we are the result of a cosmic accident, there is absolutely no reason to believe we would be located in such a special place, even if it exists. So, Big Bang theorists reject this possibility and claim that hypothetical observers in other galaxies would *also* see galaxies receding away from them. Of course, that's an unverifiable assumption. But for Christians, it makes perfect sense that both Earth and our galaxy might be in a special place, since "this is the location God chose to situate His image in man."[20]

Even if galaxy redshifts are due to an expansion, this doesn't necessarily imply that the universe is expanding today. What we believe about the time it

Figure 3. *The first JWST deep field image of distant galaxies*

takes distant starlight to reach us critically influences the way we interpret astronomical data. Because time and space are intertwined in Einstein's theory of relativity, it appears impossible to calculate the time for distant starlight to reach us without making unproven (and possibly unprovable) assumptions.[1] Evolutionary scientists assume distant light requires billions of years to reach us, and this seems reasonable in their atheistic worldview. On the other hand, biblical creationists assume that God got the light to us quickly, because He stated it was His intent to do so (Genesis 1:14–15).

In one recent creation cosmology, God caused the size of the universe to rapidly expand during Days 2 through 4 of the creation week. This expansion then slowed, and perhaps even stopped, afterward.[21] Light emitted on Day 4 by these newly created galaxies would be redshifted if we are seeing light that was emitted before God slowed or stopped the expansion. So, depending on the means God used to get distant starlight to us quickly, these redshifts could be evidence of a past expansion of the universe, even if that expansion is no longer occurring today.

Conclusion

In any case, data and images from the James Webb Space Telescope (Figure 3) are proving enormously problematic for advocates of the Big Bang. Bible-believing Christians can be confident the heavens will continue to testify to their supernatural creation by the Lord Jesus Christ. As the late ICR President Emeritus Dr. John Morris liked to say, "It's a great time to be a Bible-believing Christian!"

References

1. Hebert, J. 2022. Seeing Distant Starlight in a Young Universe. *Acts & Facts*. 51 (7): 13–15.
2. Hebert, J. 'Early' Spiral Galaxy Surprise. *Creation Science Update*. Posted on ICR.org May 1, 2020, accessed August 22, 2022.
3. Harrison, M. Scientists Puzzled Because James Webb is Seeing Stuff That Shouldn't Be There. *The Byte*. Posted on futurism.com August 30, 2022, accessed August 30, 2022.
4. Witze, A. Four revelations from the Webb telescope about distant galaxies. *Nature*. Posted on nature.com July 27, 2022, accessed August 31, 2022.
5. Bakich, M. E. The Beginning to the End of the Universe: The first stars are born. *Astronomy*. Posted on astronomy.com January 13, 2021, accessed September 1, 2022.
6. An official NASA Big Bang timeline image from over a decade ago (and archived at Wikimedia Commons) states the first stars formed 400 million years ago.
7. Schilling, G. Webb Telescope Shatters Distance Records, Challenges Astronomers. *Sky & Telescope*. Posted on skyandtelescope.org August 10, 2022, accessed August 25, 2022.
8. Hebert, J. and B. Thomas. 2014. Does Science Support the Big Bang? *Acts & Facts*. 43 (7): 21.
9. Hebert, J. 2018. Does the Cosmic Microwave Background Confirm the Big Bang? *Acts & Facts*. 47 (6): 10–12.
10. Cooper, K. The James Webb Space Telescope never disproved the Big Bang. Here's how that falsehood spread. *Space.com*. Posted on space.com September 7, 2022, accessed September 7, 2022.
11. Humorously, after Kirkpatrick's self-doubt of her life's work was publicized, she wrote "Allison the Big Bang happened Kirkpatrick" on her Twitter account!
12. Disney, M. J. AT THE BEGINNING OF TIME; what the James Webb telescope might see. Posted on mjdisney.org July 21, 2022, accessed September 1, 2022.
13. Tolman, R. C. 1930. On the Estimation of Distances in a Curved Universe with a Non-Static Line Element. *Proceedings of the National Academy of Sciences*. 16 (7): 511–520.
14. That is, assuming the test is performed using what astronomers call AB magnitudes.
15. Disney, M. J. and R. H. Lang. 2012. The galaxy ancestor problem. *Monthly Notices of the Royal Astronomical Society*. 426 (3): 1731–1749.
16. Sandage, A. 2010. The Tolman Surface Brightness Test for the Reality of

the Expansion. V. Provenance of the Test and a New Representation of the Data for Three Remote *Hubble Space Telescope* Galaxy Clusters. *The Astronomical Journal.* 139 (2): 728–742.

17. Lerner, E. J. 2018. Observations contradict galaxy size and surface brightness predictions that are based on the expanding universe hypothesis. *Monthly Notices of the Royal Astronomical Society.* 477 (3): 3185–3196.

18. Koberlein, B. The Latest Webb Observations Don't Disprove the Big Bang, But They Are Interesting. *Universe Today.* Posted on universetoday.com August 23, 2022, accessed September 26, 2022.

19. Hawking, S. 1998. *A Brief History of Time*, 10th anniversary ed. New York: Bantam Books, 44–45.

20. Morris, J. 2007. Earth: A Special Place. *Acts & Facts.* 36 (11): 13.

21. Humphreys, D. R. 2022. A more biblical cosmology. *Journal of Creation.* 36 (3): 1–9.

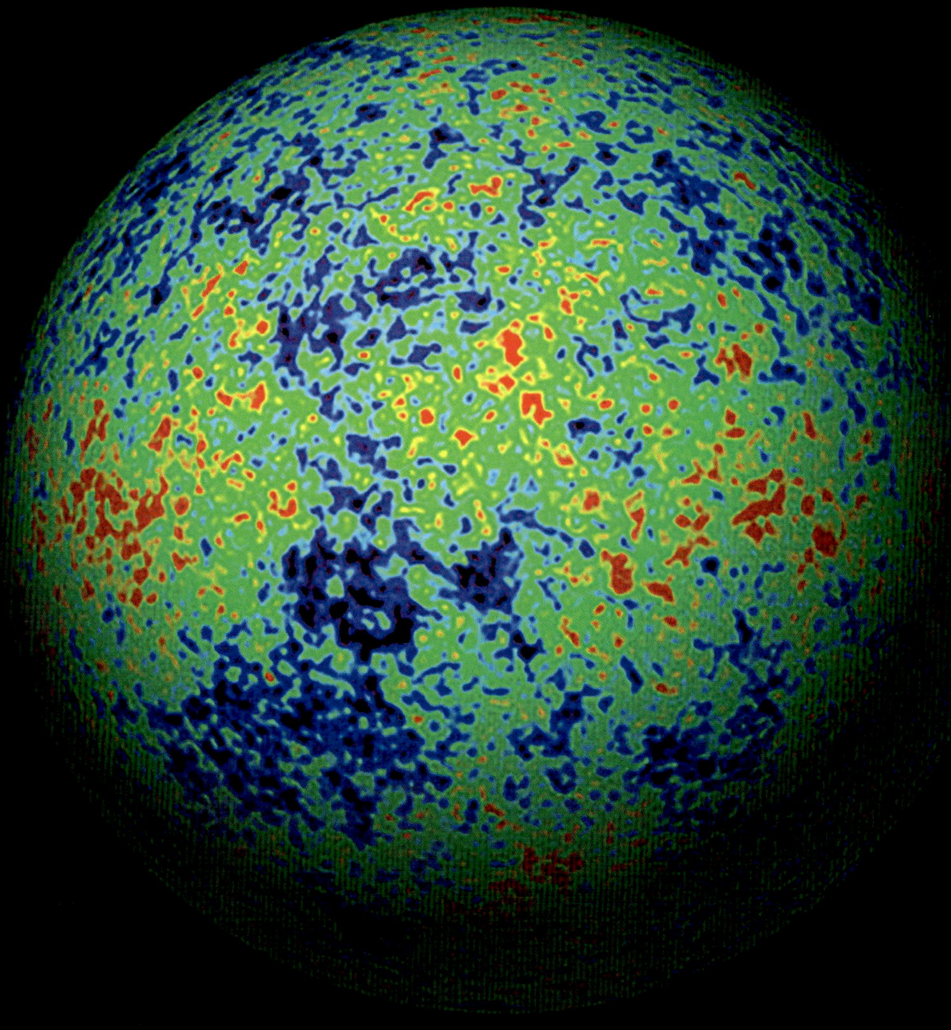

Wilkinson Microwave Anisotropy Probe (WMAP) satellite map of the CMB temperature

10
DOES THE COSMIC MICROWAVE BACKGROUND CONFIRM THE BIG BANG?

Jake Hebert, Ph.D.

Three main arguments are commonly used to support the Big Bang model of the universe's origin:

1. the apparent expansion of the universe, inferred from redshifted spectra of distant galaxies;
2. the Big Bang can account for the observed relative abundance of hydrogen and helium;
3. and the observed cosmic microwave background (CMB) radiation, thought to be an "afterglow" from a time about 400,000 years after the supposed Big Bang.

Although an expanding universe is consistent with the Big Bang, it doesn't necessarily demand a big bang as its cause. One could imagine that for some reason God imposed an expansion on His created universe, perhaps to keep the universe from collapsing under its own gravity. Of course, this assumes that conventional scientists' interpretation of the redshift data is correct, which some creation scientists are starting to question.[1]

The second argument isn't as impressive as it sounds. The Big Bang model is able to account for the observed abundance of hydrogen and helium because it contains an adjustable parameter called the baryon-to-photon ratio. Big Bang scientists simply choose a value for this parameter that gives them the right answer. Even so, it's not clear that the Big Bang can account for the total number of atoms (per unit volume) in the universe. And even with this adjustable parameter, the Big Bang cannot correctly account for the observed amounts of lithium isotopes.[2,3]

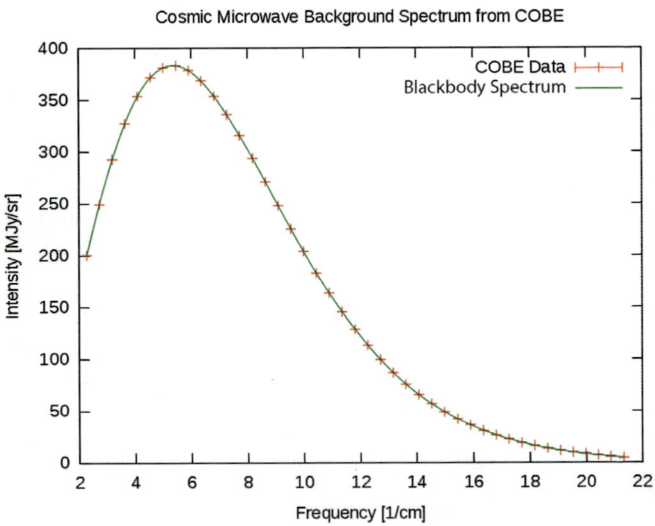

Figure 1. *The intensity of the CMB radiation (as a function of frequency) very closely matches the intensity profile of a blackbody having a temperature of 2.7 kelvins. Measurements were taken by NASA's Cosmic Background Explorer (COBE) satellite.*

However, the third argument, the existence of the CMB radiation, was successfully predicted by the Big

Bang. We observe very faint but uniform electromagnetic radiation—radiation not associated with particular stars or galaxies—coming from all directions in space, and the intensity of this radiation is brightest in the microwave part of the electromagnetic spectrum. Big Bang scientists interpret this to be the oldest light in the universe, light emitted when the universe became cool enough for neutral hydrogen atoms to form. As the universe expanded, the wavelengths of these traveling photons were stretched so that most of them had wavelengths corresponding to the microwave part of the electromagnetic spectrum. The intensity of this CMB radiation (as a function of wavelength or frequency) very closely matches the intensity of the radiation given off by an ideal emitter/absorber that physicists call a blackbody. Such a blackbody would have a temperature of 2.7 Kelvins, or about -270° Celsius (Figure 1).

The CMB and Inflation Theory

Despite successfully predicting the CMB's existence, the Big Bang has often been wrong about the CMB's details. To understand why, let's review some of the Big Bang's main assumptions.

The Big Bang model assumes there are no special places or directions in space. In other words, it supposes the universe is homogeneous and isotropic. Because the Big Bang assumes there are no special directions in space, any direction should look much the same as any other direction. And because space is supposed to look nearly the same in all directions, the CMB should *also* look nearly the same in any direction. This is indeed the case. So far, so good.

But Big Bang theorists soon realized that the uniformity of the CMB, even though required by their model, was also a problem for the Big Bang model. Such uniformity requires all points within the supposed primeval fireball to be characterized by the same temperature. Given that the Big Bang was supposedly undirected and purposeless, this requires very unlikely initial conditions. Generally, these theorists don't like this because it could be construed as fine-tuning, or evidence for a Designer. However, they could also explain the uniformity if radiant energy from one direction in space "warmed up" space in another direction. But this radiant energy travels at the finite (but still very fast) speed of light, and even the 13.8 billion years allowed by the Big Bang model is not enough time for this process to completely equalize the CMB everywhere in the universe. This is known as the horizon problem.[4]

So, theorists proposed that very shortly after the Big Bang the universe underwent a huge growth spurt called inflation, in which space itself expanded faster than the speed of light. Supposedly, the part of the universe we see was, before inflation, small enough that traveling radiant energy *did* have sufficient time to warm the cool spots. Inflation would have "ballooned" the universe so much that the part of the CMB we can observe is very uniform—as demanded by Big Bang assumptions.

Not surprisingly, there is no direct evidence for inflation. What was believed to be evidence for inflation made front-page headlines in 2014 but was quickly retracted.[5] Moreover, inflation theory has become so strange that even conventional cosmologists harshly criticize it.[6]

A CMB Surprise

According to the Big Bang model, however, the CMB should not be *perfectly* uniform. Although the Big Bang assumption of isotropy does not allow for temperature differences that stretch across large patches of the sky, it predicts that small patches (with an angular diameter of about 1° or less) should show extremely subtle variations in temperature, with some spots being *very* slightly warmer or cooler than 2.7 Kelvins.[7] Conventional cosmologists originally expected the sizes of these temperature differences to be about one ten-thousandth of a Kelvin, but later measurements showed that this expectation was 10 times too large; the actual measured differences were about a hundred-thousandth of a Kelvin.[8]

Creation astronomer Dr. Danny Faulkner often recounts his personal memory of how surprised conventional scientists were when the measured temperature differences were smaller than expected. So, despite popular perception, observations have *not* always agreed with Big Bang expectations. Instead, the Big Bang has often been tweaked to absorb anomalous observations.

Challenges for Conventional and Creation Scientists

Even so, there are other problems with the Big Bang interpretation of the CMB. Although small-

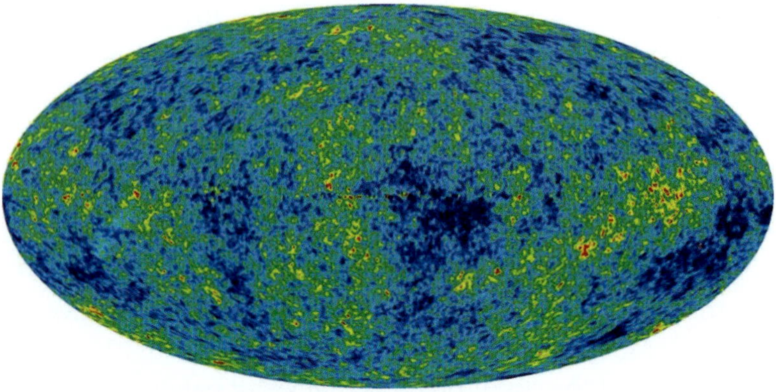

Wilkinson Microwave Anisotropy Probe (WMAP) map of the cosmic microwave background

scale (angular diameter of less than 1°) temperature anisotropies (differences) are expected in the Big Bang model, there also exist very subtle temperature anisotropies within the CMB stretching across *large* patches of the sky. Because the Big Bang assumes isotropy, these large-scale anisotropies should not exist. Conventional scientists had hoped that better data would remove these apparent anomalies, but higher-resolution CMB data obtained by the Planck satellite disappointed them. George Efstathiou, a Cambridge astrophysicist and one of the leaders of the Planck satellite's science team, explained in an interview:

> The theory of inflation predicts that today's universe should appear uniform at the largest scales in all directions....That uniformity should also characterize the distribution of fluctuations at the largest scales within the CMB. But these anomalies, which Planck confirmed, such as the cold spot, suggest that

> this isn't the case....This is very strange....And I think that if there really is anything to this, you have to question how that fits in with inflation....It's really puzzling.[9]

To summarize, the Big Bang assumes that the universe is isotropic—space should basically look the same in all directions. Since the universe is assumed to be isotropic, the CMB should also be isotropic. And the CMB *is* very isotropic, which would seem to be a plus for the Big Bang model. But Big Bang cosmologists realized that isotropy was extremely unlikely, so they proposed inflation to explain *why* the CMB was isotropic. But there is no evidence supporting inflation theory, and the theory has become so weird that it's now being harshly criticized by other conventional scientists. The small-scale temperature differences in the CMB were about 10 times smaller than predicted, so the Big Bang was tweaked to bring these temperature differences into agreement with the model. But the high-resolution Planck data show that subtle large-scale features in the CMB *are* present—features that are not supposed to exist if inflation is correct!

But it gets worse. Some of these large-scale temperature differences in the CMB seem to be aligned with our solar system.[10] As acknowledged by outspoken Big Bang proponent Lawrence Krauss:

> But when you look at CMB map, you also see that the structure that is observed, is in fact, in a weird way, correlated with the plane of the earth around the sun. Is this Copernicus

> coming back to haunt us? That's crazy. We're looking at the whole universe. There's no way there should be a correlation of structure with our motion of the earth around the sun—the plane of the earth around the sun—the ecliptic. That would say we are truly the center of the universe.[11]

This apparent alignment, if real, is a huge problem for the Big Bang since it seems wildly improbable that Earth should be in a special place in the universe. Don't forget that according to the Big Bang, special places and directions aren't supposed to exist. Creation scientists have no problem with Earth being near the center of the universe, since this might be expected given God's great care and attention for this planet.

But the CMB also presents challenges for creation scientists because it needs to be explained within a creation context. Recently, Dr. Faulkner and creation physicist Dr. Russell Humphreys both proposed possible explanations for the CMB within a biblical worldview.[12,13] However, more work needs to be done to refine and decide between competing models.

Conclusion

The successful prediction of the existence of a cosmic microwave background is arguably the Big Bang's greatest success, and creation scientists need to provide a detailed alternative explanation for its existence. Even so, there are features within the CMB that are completely contrary to Big Bang expectations. Given that the Big Bang's *strongest* argument has these problems, one can safely conclude that the scientific case for the Big Bang model is extremely weak, and no one, especially Christians, should feel compelled to accept it.

References

1. Hartnett, J. 2011. Does observational evidence indicate the universe is expanding?—part 2: the case against expansion. *Journal of Creation.* 25 (3): 115–120.
2. Hebert, J. 2013. Dark Matter, Sparticles, and the Big Bang. *Acts & Facts.* 42 (9): 17–19.
3. Thomas, B. Big Bang Fizzles under Lithium Test. *Creation Science Update.* Posted on ICR.org September 22, 2014, accessed March 7, 2018.
4. Coppedge, D. F. 2007. The Light-Distance Problem. *Acts & Facts.* 36 (6).
5. Hebert, J. Big Bang Evidence Retracted. *Creation Science Update.* Posted on ICR.org February 12, 2015, accessed March 9, 2018.
6. Hebert, J. Big Bang Blowup at Scientific American. *Creation Science Update.* Posted on ICR.org May 29, 2017, accessed March 9, 2018.
7. In the 13.8 billion years allowed by the Big Bang model, traveling radiation would only have had sufficient time to equalize the CMB in small patches of the sky (with angular diameters less than about 1°).
8. Wright, E. L. 2004. Theoretical Overview of Cosmic Microwave Background Anisotropy. In *Measuring and Modeling the Universe*, Carnegie Observatories Astrophysics Series, vol. 2. W. L. Freedman, ed. Cambridge, UK: Cambridge University Press, 292.
9. Discoveries from Planck may mean rethinking how the universe began. The Kavli Foundation. Posted on phys.org July 26, 2013, accessed March 9, 2018.
10. Huterer, D. 2007. Why is the solar system cosmically aligned? *Astronomy.* 35 (12): 38–43.
11. The Energy of Empty Space That Isn't Zero: A Talk with Lawrence Krauss. *Edge.* Posted on edge.org July 5, 2006, accessed March 9, 2018.
12. Faulkner, D. R. 2016. Thoughts on the *raqîa'* and a Possible Explanation for the Cosmic Microwave Background. *Answers Research Journal.* 9: 57–65.
13. Humphreys, D. R. 2014. New view of gravity explains cosmic microwave background radiation. *Journal of Creation.* 28 (3): 106–114.

11
DARK MATTER SEARCH KEEPS COMING UP EMPTY

Jake Hebert, Ph.D.

A 2020 study ruled out a possible candidate for dark matter, the mysterious invisible "stuff" said to comprise 85% of the matter in our universe.[1,2] The Big Bang model needs dark matter for a number of reasons. So, this is bad news for the Big Bang model.

Most conventional cosmologists believe in dark matter. Some biblical creationists also think dark matter exists,[3] but others (including me) are more skeptical.[4] Scientists, both creationist and conventional, who are skeptical of dark matter tend to think that what is being construed as evidence for it is really due to an incomplete understanding of how gravity behaves over very large distances.[5] Creationists who think dark matter is real point out that dark matter was not invented simply to "fix" problems with the Big Bang—astronomers were discussing the possibility of dark matter at about the same time and even before the Big Bang model was first proposed.

Although this is true, dark matter has nevertheless become extremely important to conventional scientists. Big Bang cosmologists now invoke dark matter in their theories of galaxy formation.[6] Furthermore, Big Bang cosmology *demands* that this dark matter be composed of exotic, never-before-detected particles. This is because Big Bang cosmologists must assume a particular value for an adjustable parameter in their model in order for the Big Bang to produce the observed abundance of helium and hydrogen. Once this is done, the Big Bang model can only generate enough "normal" matter (atoms) to account for about 15% of the matter thought to exist. This means that, by their reckoning, the dark matter can't be made of atoms. But they have ruled out other forms of non-atomic matter (neutrinos, free electrons, etc.) for other reasons. This means that some exotic, new form of matter is the only remaining option.[7]

But the search for this exotic matter is not going very well. In one disappointment, astronomers had speculated that an X-ray "glow" around distant clusters of galaxies was caused by the decay of a particular dark matter candidate called a sterile neutrino. If this were the case, then this X-ray glow should surround our own galaxy since they think the Milky Way has its own dark matter halo. But this X-ray glow is absent from the Milky Way.[1,2] Astroparticle physicist Ben Safdi (University of Michigan, Ann Arbor) said, "It seems to be the final nail in the coffin of the dark matter interpretation [of the X-ray glow]."[2]

Big Bang theorists are becoming increasingly desperate to find this "missing" dark matter. Some have now speculated that dark matter perhaps pre-dated the Big Bang.[8]

Given the continuing woes besetting the Big Bang model, it is a wonder that so many, including so many evangelical Christians, seem to accept this model despite its enormous scientific problems and its clear contradiction with Scripture.[9,10]

References

1. Desert, C. et al. 2020. The dark matter interpretation of the 3.5-keV line is inconsistent with blank-sky observations. *Science*. 367 (6485): 1465–1467.
2. Grossman, L. A controversial X-ray glow didn't show up in the Milky Way's dark matter halo. *Science News*. Posted on sciencenews.org March 26, 2020, accessed April 2, 2020.
3. Faulkner, D. 2017. The Case for Dark Matter. *Answers Research Journal*. 10: 89–101.
4. Coppedge, D. F. Will Dark Matter Hunters Ever Give Up? *Creation Evolution Headlines*. Posted on crev.info April 1, 2020, accessed April 3, 2020.
5. Letzter, R. 19 Galaxies Are Apparently Missing Dark Matter. No One Knows Why. *Space.com*. Posted on space.com December 5, 2019, accessed April 3, 2020.
6. Assembly of Galaxies. James Webb Space Telescope/Goddard Space Flight Center website. Accessed April 3, 2020.
7. Hebert, J. 2013. Dark Matter, Sparticles, and the Big Bang. *Acts & Facts*. 42 (9): 17–19.
8. Hebert, J. 2019. Continuing Troubles for the Big Bang Model. *Acts & Facts*. 48 (11): 10–13.
9. Thomas, B. and J. Hebert. 2014. Does Science Support the Big Bang? *Acts & Facts*. 43 (7): 21.
10. Morris, J. D. 1997. Is the Big Bang Biblical? *Acts & Facts*. 26 (5).

12
DOES THE UNIVERSE LOOK OLD?

Jake Hebert, Ph.D.

Since distant galaxies are billions of light-years away, some understandably assume that distant starlight must have taken billions of years to reach us. Also, some galaxies are colliding. While God could have certainly created galaxies in the process of colliding, it's easy to see, given the vast distances between galaxies, how some could interpret this as evidence of great age.

However, the reality is more nuanced. The main argument for an old earth and solar system comes from radioisotope dating methods. Yet, these methods and ages are highly problematic (such as wrong dates given to rocks of known ages). Other than that, our solar system really does look young. Practically everywhere we look in our solar system we see strong indications of youth, including "upper limits" for ages of solar system objects that are measured in just thousands of years.[1]

The deep-space argument for an old universe seems stronger, but there is much we don't know about cosmology and the universe, and such arguments often involve unproven—and, in some cases, unprovable—assumptions.[2] Even so, there are hints that something is wrong with the conventional view. If the universe really is billions of years old and the Big Bang model is correct, then we are observing distant galaxies not as they are today but as they were billions of years ago. This means that these galaxies should appear undeveloped and "unevolved." Yet distant galaxies routinely contradict that expectation.[3] This means there is something wrong with either conventional ideas of galaxy evolution or the assumption that distant starlight takes billions of years to reach us or both.

Moreover, some deep-space objects seem far younger than they should be if the universe were 13.8 billion years old. Spiral galaxies should have lost their spiral structure long ago, and conventional astronomers have been repeatedly surprised by youthful features in globular clusters.[4] Some exoplanets are so close to their host stars that they should no longer exist after billions of years![5]

Albert Einstein

Many creationists think Einstein's theory of relativity, which allows clocks to tick at different rates, is the key to resolving these mysteries. God might have allowed clocks in deep space to tick at much faster rates in order to let distant starlight reach us during the literal 24 hours of the creation week's Day 4.

It is noteworthy that the strongest arguments for an old universe, such as distant starlight, involve extremely distant objects and unproven assumptions. Yet within our solar system, where we have a wealth of up-close observations, evidence for youth is both straightforward and strong. Coincidence? Or a hint that creationists have an edge when interpreting astronomical data?

Although mysteries remain, Christians have strong reasons, both biblical and scientific, to affirm that the Lord Jesus Christ created the entire universe just a few thousand years ago.

References

1. Hebert, J. 2018. Our Young Solar System. *Acts & Facts.* 47 (9): 10–13.
2. Arledge, S. 2021. Teachable Minds and Scientific Discovery. *Acts & Facts.* 50 (2): 20.
3. Hebert, J. Most Distant Rotating Disk Galaxy Challenges Secular Models. *Creation Science Update.* Posted on ICR.org May 28, 2020, accessed August 6, 2021.
4. Hebert, J. 2019. Deep-Space Objects Are Young. *Acts & Facts.* 48 (9): 10–13.
5. Hebert, J. Hot Neptune Atmosphere "Shouldn't Exist." *Creation Science Update.* Posted on ICR.org November 5, 2020, accessed August 6, 2021.

13
DEEP-SPACE OBJECTS ARE YOUNG

Jake Hebert, Ph.D.

Conventional astronomers claim our universe is unimaginably ancient—almost 14 billion years old. Yet the Bible clearly teaches that God created the universe in the relatively recent past, about 6,000 years ago. A 2018 *Acts & Facts* article described how observations in our own solar system are much more consistent with recent creation than with long ages.[1] But what about observations *beyond* our solar system? Are deep-space observations more consistent with a young universe or an old one?

"Winding Up" of Spiral Galaxies

Spiral galaxies are so named because of their beautiful spiraling arms. A spiral galaxy slowly spins around an axis of rotation perpendicular to its disk. Stars, dust, and gases closer to the axis of rotation take less time to complete a revolution than material that is farther out. As a result, the spiral arms "wind up," and eventually the spiral shape becomes unrecognizable. Back-of-the-envelope calculations and simplified models suggest that a few hundred million years of additional winding would cause the spiral

Figure 1. *The Whirlpool galaxy is a well-known spiral galaxy*

structure to become noticeably tighter than what we observe. Some astronomers claim the spiral structure would be *completely destroyed* in a few hundred million years.[2] In any case, winding would certainly have obliterated the spiral structure after five billion years or so (Figure 2). Yet conventional scientists claim that many spiral galaxies, including our own Milky Way galaxy, are around 10 billion years old. If this is the case, then why are they still recognizable as spiral galaxies?

This argument assumes that the spiral arms are physical objects and that the same material remains within a particular spiral arm over time. Conventional scientists argue this isn't necessarily the case. They claim spiral arms are not real physical structures but

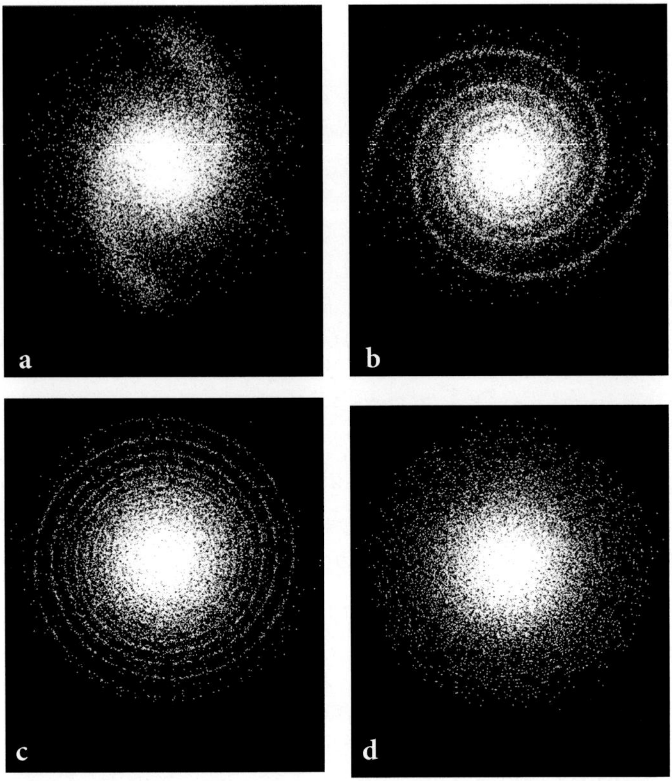

Figure 2. *This shows the results of a simplified computer model of spiral galaxy winding. Assuming the spiral galaxy depicted in a) has an age of zero, the degree of winding in b) is after 300 million years and in c) is after one billion years. d) After five billion years, the spiral structure has been erased.*

are instead high-density areas that rotate around the galaxy at a different rate than the disk materials. These high-density areas are similar to obstructions or "clogs" of cars in a traffic jam. These traffic obstructions persist over time, even though different individual cars are continually passing into and out of the "clog." In a similar fashion, conventional scientists would argue that different stars, gases, and dust particles are continually moving through the high-density regions. Supposedly, this allows the spiral structure to persist for billions of years. It also supposedly allows the formation of new stars.[2]

However, it's premature to say that this "spiral density wave" model solves the problem. It was proposed in the 1940s and refined in the 1960s, but conventional scientists are still working on it.[3,4] The results of computer simulations have been inconsistent, with some seeming to confirm the theory[5] and others appearing to contradict it.[6,7] One such simulation suggested that the stars really do rotate with the spiral arms after all.[7]

Conventional astronomers have admitted the theory has difficulties. What causes the density wave in the first place?[8,9] What maintains the wave over time?[8] According to an online astronomy encyclopedia:

> The density wave model is a work in progress, with one of the big remaining questions being how the density waves survive for such long periods. Given the enormous amount of energy required to compress the interstellar gas and dust, one would expect them to die away over time.[10]

For these reasons, spiral galaxies are a valid young-universe argument, regardless of conventional objections.

Hot Blue Stars

High-mass stars are hotter and bluer than less massive stars, and they consume their nuclear fuel much more quickly. By conventional reckoning, a hot blue star with a mass 15 times greater than our sun has a lifetime of just 15 million years (Figure 3). The very hottest blue stars, with masses around 25 times greater than our sun, have lifetimes of just a few million years.[11] Creation astronomer Ron Samec pointed out that these estimated lifetimes are based on computer models of stellar evolution. These models are calibrated by assuming that our sun is 4.6 billion years old.[12]

Yet nothing requires scientists to make this assumption. Famed solar astronomer John Eddy once stated:

Figure 3. *The Pleiades star cluster is known for its beautiful blue stars. The very hottest blue stars have lifetimes (based on uniformitarian assumptions) of just a few million years.*

> I suspect…that the sun is 4.5 billion years old. However, given some new and unexpected results to the contrary, and some time for frantic recalculation and theoretical readjustment, I suspect that we could live with Bishop Ussher's value [about 6,000 years] for the age of the earth and sun. I don't think we have much in the way of observational evidence in astronomy to conflict with that.[13]

Creation scientists argue that the sun and stars are truly just thousands of years old. So, these estimated star lifetimes, obtained by assuming our sun is 4.6 billion years old, could have large errors. In any case, even by conventional reckoning hot blue stars can only last millions of years before they eventually explode. So, if the universe is 13.8 billion years old, why do hot blue stars still exist?

Conventional scientists claim new stars are born to replace old blue stars that have died. The spiral density wave model supposedly explains the continued presence of blue stars in spiral galaxies. But, as noted earlier, this model has problems. And what about blue stars found in places other than spiral galaxies?

Furthermore, naturalistic star formation, though perhaps barely possible, is extremely unlikely. Astrophysicist Neil deGrasse Tyson, no friend of the creation movement, once remarked, "The scary part is that if none of us knew in advance that stars exist, front line research would offer plenty of convincing reasons why stars could never form."[14]

One of these reasons is that conventional star formation theories require at least one generation of stars to already be in existence.[15] But a theory of star formation that needs pre-existing stars is not very compelling. The very existence of billions of stars in our universe is a silent rebuke to the Big Bang model, and the continued existence of hot blue stars is an argument that our universe is young.

Globular Clusters

Globular clusters are beautiful, spherical collections of stars that orbit their host galaxies. Our Milky Way galaxy contains about 150 known globular clusters (Figure 4). By conventional reckoning, globular clusters are around nine to 10 billion years old—or even older.[16]

Neutron stars within globular clusters are a problem for old-universe beliefs. As the name implies, a neutron star is composed of subatomic particles called neutrons. A typical neutron star may have a mass about one-and-a-half times greater than our sun, but this material is squeezed into a body only 10 miles across. This means that neutron stars are incredibly dense: a tablespoon of neutron star material would have the same mass as Mt. Everest![17]

Neutron stars are thought to be formed in certain supernova explosions. Neutron stars receive a noticeable "kick" from the explosions, giving them speeds very different from their "parent" stars. These speeds can be hundreds or even thousands of kilometers per second.

The higher a neutron star's speed, the easier it

***Figure 4**. Messier 80, one of the densest of the Milky Way's 150 or so globular star clusters*

can escape the cluster's gravitational pull. In fact, the speeds of most neutron stars are so high that they should be able to escape from globular clusters in just thousands of years. Furthermore, high-mass stars (like those in globular clusters) are supposed to generate neutron stars fairly quickly. Since globular clusters are supposedly billions of years old, few if any neutron stars should still remain within globular clusters. Yet some Milky Way globular clusters contain hundreds or even thousands of such neutron stars.[18] This neutron star retention problem strongly suggests that globular clusters are just thousands, not billions, of years old. The presence of black holes within globular clusters may also be a problem for deep time for a similar reason.[19,20]

Furthermore, the innermost stars within some globular clusters are rotating around a central axis. Yet gravitational interactions between stars should have long ago erased any such rotational tendency.

> Theory and numerical simulations of globular clusters indicate that any central rotation should be erased on relatively short timescales. Because these globular clusters were formed billions of years ago, we would expect that any rotation signature would have been eradicated by now.[21]

Nevertheless, many globular clusters show this rotational tendency. Perhaps globular clusters aren't billions of years old after all.

What about Distant Starlight?

Distant galaxies are millions and even billions of light-years away from us. Since the speed of light is finite, conventional scientists argue that distant starlight must take millions and billions of years to reach us.

Creation scientists disagree. The Hebrew wording used in Genesis demands that the light from stars was emitted on Day 4 of the creation week and that it also arrived that same day.[22] Perhaps this was a bona fide miracle, or perhaps the laws of physics God established allow distant light to naturally reach us in a short time. If the latter, then the solution almost certainly involves Einstein's theory of relativity, and creation scientists have proposed a number of suggestions as to how God may have done this. It should be noted that the Big Bang model has its own version of this problem, which is one of the main reasons conventional cosmologists added inflation theory to the model.[23] But inflation theory has become so strange that even conventional scientists have harshly criticized it.[24]

Conclusion

Other deep-space objects pose problems for deep time. Conventional scientists claim that distant supermassive black holes formed less than a billion years after the supposed Big Bang, but they have difficulty explaining how such enormous objects could naturally form so relatively quickly.[25] Transitory structures in the Eagle Nebula's Pillars of Creation are so short-lived that conventional scientists think it is an amazing coincidence we are able to observe them today.[26]

Despite the dogmatic assertions of conventional scientists, their claim that the universe is billions of years old is not a foregone conclusion. Challenges confront both the creation and conventional models, but the preponderance of evidence clearly favors recent creation.

References

1. Hebert, J. 2018. Our Young Solar System. *Acts & Facts*. 47 (9): 10–13.
2. Freedman, R. A. and W. J. Kaufmann III. 2002. *Universe: Stars and Galaxies*. New York: W. H. Freeman and Co., 576–579.

3. Lin, C. C. and F. H. Shu. 1964. On the spiral structure of disk galaxies. *Astrophysical Journal*. 140: 646–655.
4. Shu, F. H. 2016. Six Decades of Spiral Density Wave Theory. *Annual Review of Astronomy and Astrophysics*. 54 (1): 667–724.
5. Peterken, T. G. et al. 2019. A direct test of density wave theory in a grand-design spiral galaxy. *Nature Astronomy*. 3: 178–182.
6. Foyle, K. et al. 2011. Observational Evidence Against Long-Lived Spiral Arms in Galaxies. *The Astrophysical Journal*. 735 (2): 101.
7. New theory of evolution for spiral galaxy arms. Royal Astronomical Society. Posted on phys.org April 21, 2011, accessed July 1, 2019.
8. Strobel, N. Nick Strobel's Astronomy Notes. Posted on astronomynotes.com, accessed July 5, 2019.
9. Mihos, C. Astronomy 222 – Galaxies and Cosmology Notes. Case Western Reserve University. Posted on burro.cwru.edu.
10. Density Wave Model. COSMOS - The SAO Encyclopedia of Astronomy. Swinburne University of Technology. Posted on astronomy.swin.edu.au, accessed July 8, 2019.
11. Freedman and Kaufmann, 481.
12. Samec, R. G. and E. Figg. 2012. The apparent age of the time dilated universe 1: gyrochronology, angular momentum loss in close solar type binaries. *Creation Research Society Quarterly*. 49 (1): 5–19.
13. Kazman, R. G. 1978. Interview with John A. Eddy. It's About Time: 4.5 Billion Years (a report on the symposium at Louisiana State University). *Geotimes*. 23: 18. Cited in Lubenow, M. 1992. *Bones of Contention*. Grand Rapids, MI: Baker Book House, 205. Emphasis in Lubenow.
14. Tyson, N. deG. 2007. *Death by Black Hole: And Other Cosmic Quandaries*. New York: W. W. Norton & Company, 187.
15. One popular explanation is that the shock from a supernova compresses interstellar gas into a high enough density for star formation. But a supernova is an exploding star!
16. Klesman, A. Globular clusters might be younger than we thought. *Astronomy*. Posted on astronomy.com June 18, 2018, accessed May 28, 2019.
17. Mikles, V. If a tablespoon of a neutron star were placed on Earth's surface, what would happen? *Astronomy*. Posted on astronomy.com August 28, 2018, accessed June 6, 2019.
18. Nethercott, P. 2016. Neutron Stars in Globular Clusters: Evidence of Young Age? *Creation Research Society Quarterly*. 53 (1): 14–18.
19. Surprising black-hole discovery changes picture of globular star clusters. National Radio Astronomy Observatory. Posted on phys.org October 3, 2012, accessed July 1, 2019.

20. Davis, J. Physicists find black holes in globular star clusters, upsetting 40 years of theory. *Phys.org*. Posted on phys.org November 4, 2013, accessed May 8, 2019.

21. Globular clusters rotate at heart. University of Texas McDonald Observatory press release. Posted on mcdonaldobservatory.org May 9, 2014, accessed May 8, 2019.

22. Johnson, J. J. S. 2018. Uniformitarians Stumble on Distant Starlight. *Acts & Facts*. 47 (1): 21.

23. Coppedge, D. F. 2007. The Light-Distance Problem. *Acts & Facts*. 36 (6).

24. Hebert, J. Big Bang Blowup at Scientific American. *Creation Science Update*. Posted on ICR.org May 29, 2017, accessed July 5, 2018.

25. Saplakoglu, Y. Zeroing in on How Supermassive Black Holes Form. *Scientific American*. Posted on scientificamerican.com September 29, 2017, accessed April 10, 2019.

26. Hebert, J. The Hubble 'Pillars of Creation' Revisited. *Creation Science Update*. Posted on ICR.org January 19, 2015, accessed July 1, 2019.

14
"EARLY" SPIRAL GALAXY SURPRISE

Jake Hebert, Ph.D.

The naked eye allows us to see just a little of God's heavenly handiwork, but even this little bit clearly declares God's glory.[1] As more powerful telescopes peer deeper into space, more and more reasons to question the conventional origins story accumulate.

Recently, astronomers were surprised by very distant spiral galaxies. As described in a popular science news article, "New results from an ambitious sky survey program, called ALPINE, reveal that rotating disk-shaped galaxies may have existed in large numbers earlier in the universe than previously thought."[2] These rotating galaxies were detected using data from multiple observatories, including Chile's Atacama Large Millimeter/submillimeter Array, or ALMA. By observing Doppler shifts[3] of light emitted from positively charged carbon ions within the galaxies, astronomers were able to infer that the galaxies were rotating. By conventional reckoning, these galaxies would have formed between one and one and a half billion years after the supposed Big Bang, or 12.2 to 12.7 billion years ago.[2]

Andromeda galaxy

Big Bang astronomers assume that light from the most distant galaxies takes billions of years to reach Earth. Indeed, this assumption is often used to challenge Bible believers. It is at the heart of the question, "If the universe is just thousands of years old, how can we see starlight from galaxies that are billions of light-years away?"

However, Einstein's theories of relativity have shown that supposedly intuitive ideas about light, space, and time are incorrect, and many creationists think that relativity theory may be the key to answering this common objection to biblical creation.

Moreover, this assumption is a double-edged sword for conventional astronomers, because distant light presents problems for the Big Bang, too. Because Big Bang astronomers assume light from the most distant galaxies takes billions of years to reach

Earth, they think we are seeing these very distant galaxies as they were in the very distant past. Since galaxies supposedly take vast amounts of time to form and "evolve," the most distant galaxies should appear immature and poorly developed to us. However, many galaxies that are extremely distant appear fully formed or mature.[4–7]

Better observations are needed to confirm if these disk-shaped objects are true spiral galaxies. If so, this would turn out to be just one more example of this phenomenon.

It should also be noted in passing that the pinwheel-shaped arms in spiral galaxies should become tightly wound after just hundreds of millions of years, so the continued existence of spiral galaxies is an argument that they are much younger than conventional scientists think.[8]

References

1. Psalm 19:1–2.
2. Clavin, W. Rotating galaxies galore: New results from ALPINE reveal what appear to be spiral galaxies in the infant universe. *Phys.org*. Posted on phys.org April 21, 2020, accessed April 22, 2020.
3. A familiar example of a Doppler shift is the increase in pitch of an approaching ambulance siren and the decrease in the siren's pitch as it moves away. Light can also be Doppler shifted: light spectra from sources that are moving away are shifted toward the "red" end of the spectrum, and spectra from approaching sources are shifted to the blue end of the spectrum.
4. Thomas, B. Distant Galaxies Look Too Mature for Big Bang. *Creation Science Update*. Posted on ICR.org November 30, 2011, accessed April 22, 2020.
5. Thomas, B. 'Old' Galaxy Found in 'Young' Part of Universe. *Creation Science Update*. Posted on ICR.org May 24, 2011, accessed April 22, 2020.
6. Thomas, B. Distant Galactic Cluster Should Not Exist. *Creation Science Update*. Posted on ICR.org May 21, 2010, accessed April 22, 2020.

7. Thomas, B. Secrets From the Most Distant Galaxy. *Creation Science Update*. Posted on ICR.org November 18, 2010, accessed April 22, 2020.

8. Hebert, J. 2019. Deep-Space Objects Are Young. *Acts & Facts*. 48 (9): 10–13.

9. *The Universe: A Journey Through God's Grand Design*. 2018. DVD. Dallas, TX: Institute for Creation Research.

10. Psarris, S. *What You Aren't Being Told About Astronomy*, vol. 1–3. DVD. Creation Astronomy Media.

11. *Guide to the Universe*. 2016. Dallas, TX: Institute for Creation Research.

12. Williams, J. 2010. *The Work of His Hands*. St. Louis, MO: Concordia Publishing House.

13. Staff Writers. 2017. *Space: God's Majestic Handiwork*. Dallas, TX: Institute for Creation Research.

14. Turner, J. 2019. *Space*. Dallas, TX: Institute for Creation Research.

15. Williams, J. 2018. *The Work of His Hands for Kids*. St. Louis, MO: Concordia Publishing House.

15
SEEING DISTANT STARLIGHT IN A YOUNG UNIVERSE

Jake Hebert, Ph.D.

Many see distant starlight as an unanswerable objection to recent creation. Both creationist and evolutionist astronomers agree that distant galaxies are billions of light-years from us.[1] A light-year is about six trillion miles, the distance traveled by light in one year. So, shouldn't it take billions of years for distant starlight to reach us?

Measuring the Speed of Light

Physicists use the letter *c* as the symbol for the speed of light. To determine *c*, very accurate clocks measure the time needed for light to bounce off a mirror and return to its starting point (Figure 1). For a distance of five meters between the light source and mirror, the measured speed is the round-trip distance of 10 meters divided by the total time for the trip. These measurements show that the speed of light, *c*, is 186,000 miles, or 300,000 kilometers, per second.

Light Is Weird

The speed of light is rather weird, and this makes

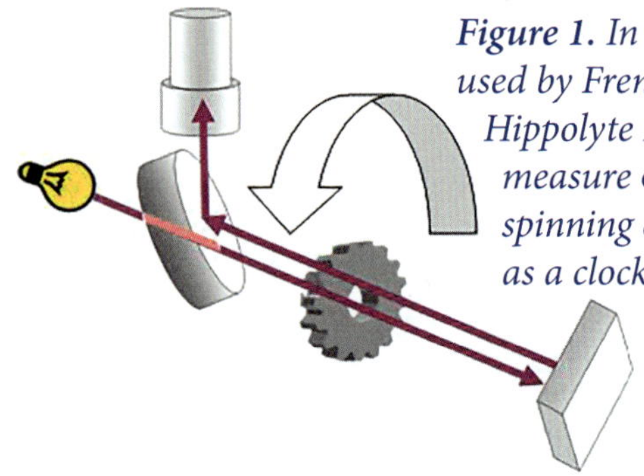

***Figure 1.** In the apparatus used by French physicist Hippolyte Fizeau to measure c, a rapidly spinning cogwheel acts as a clock.*

the issue more complicated than one might assume. The value of *c* is always 186,000 miles per second regardless of whether the light source is moving toward you, away from you, or standing still. If you stop and think about it, that's really counterintuitive. This is one of the details that led to Albert Einstein's theory of relativity, which allows for weird effects like clocks ticking at different rates.

It's ironic that episode four of the 2014 reboot of the science television series *Cosmos* explained how relativity theory invalidates our commonsense understanding of distance and time. However, that same episode used a commonsense understanding of distance and time to claim distant starlight rules out recent creation![2]

The Big Bang Model Has a Light/Time Problem

The Big Bang model has its own version of a distant starlight dilemma: the horizon problem.[3] It was one of several reasons Big Bang theorists tacked on a process called inflation to the Big Bang story.[4] Inflation solves those problems, but it creates others that are arguably even worse. Former proponents now harshly criticize inflation, claiming it isn't even a scientific theory.[5] Purported "smoking gun" evidence for inflation in 2014 was quickly retracted.[6]

Moreover, distant galaxies often seem more mature than would be expected by Big Bang reckoning. If one assumes distant starlight really does take billions of years to reach us, we should be seeing the most distant galaxies not as they are now but as they were more than 13 billion years ago, shortly after the supposed Big Bang. These most distant galaxies should look immature and "unevolved," yet they routinely contradict this expectation.[7]

Conventional astronomers therefore claim galaxy evolution somehow occurs much faster than they thought. But this assertion does nothing to explain how galaxy evolution could occur, let alone so quickly, nor why their previous expectations were so wrong.

These observations mean that either there's something wrong with evolutionary ideas about galaxy evolution or the assumption that distant light takes billions of years to reach us is mistaken—or both.

Supernatural or Natural Explanation?

Some might worry creationists are treading on "holy ground" by even attempting to explain distant starlight. The creation week was a time of unique supernatural activity, and we can't expect to understand everything about it. In fact, we can't expect to understand most of it! Even if we don't know how it worked for certain, exploring the distant starlight question is worthwhile.

As discussed below, it's at least theoretically possible that distant starlight might still be reaching us quickly even today. If that were the case, the correct explanation would not be a creation week miracle, since God finished His creative work by the end of Day 6 (Genesis 2:1–2). In that case, the explanation would be understandable in terms of normal physics.

Even if God used a miracle to get starlight to us quickly, we may be able to understand the effects of the miracle. For instance, if God decreed that *c* had a much higher value during the creation week, this is simply a miracle we can't explain scientifically. But such a miracle might have left behind astronomical clues we can explain better than Big Bang advocates can. Hence, it's useful from a theological perspective to try to understand how God brought distant starlight to Adam's eyes, even if He used a miracle to do so.

Can We Even Talk about a Distant Light Problem?

Remember that physicists measure *c* by bouncing light off a mirror and measuring the total round-trip time. You might wonder "Why do you need a mirror? Why not just use two clocks separated by some distance?"

The speed of light is so fast that you'd need precisely synchronized clocks to keep from getting a wildly erroneous answer. However, according to well-established relativity theory, even if two clocks next to one another start out synchronized, the very act of moving one of the clocks causes them to become unsynchronized. Despite many clever attempts, no one has ever figured out a way to get around this difficulty. Apparently, it's impossible to measure the one-way speed of light.

But if measuring the one-way speed of light is impossible, then it's also impossible to calculate the time for distant starlight to reach us. That travel time is apparently unknowable. And if this travel time is unknowable, can we even talk about a light travel time problem? Is such a problem even real?

Even conventional physicists have acknowledged it's theoretically possible that light from distant stars could reach us in no time at all provided the outbound speed of light is $c/2$.[8] This would imply we're seeing the universe in real time, with no time lag.

Relativity Theory to the Rescue?

As the above discussion suggests, most creationists think Einstein's theory of relativity is the key to answering this question.[9] They've generally taken two approaches to this issue. One involves our inability to measure the one-way speed of light. This inability is closely related to the fact that according to relativity theory, events seen by one observer as happening at the same time don't necessarily have to be seen as simultaneous by another observer.

A second approach, first proposed by physicist Dr. Russell Humphreys,[10] makes use of something called time dilation. Both theory and experiment indicate that clocks can tick at different rates due to differences in gravity. To put it simply, gravity makes clocks tick slower. This raises the possibility that clocks in deep space could tick much faster than

Figure 2. Given the great distances and slow relative speeds between them, collisions between galaxies seem to require millions of years. This suggests that clocks in deep space have ticked much, much faster than clocks on Earth.

clocks on Earth, allowing billions of years since creation to elapse in deep space while only 6,000 years are measured by clocks on Earth.[11]

These approaches have their pros and cons. As discussed below, time dilation seems better able to explain deep-space observations that imply vast ages, such as colliding galaxies (Figure 2). But other creationists question whether Humphreys' original time dilation model is capable of providing the billions of years of gravitational time dilation needed for it to work.[12]

An infinitely fast inbound speed of light is an easy-to-understand way to answer the distant starlight objection, but many see it as philosophically and scientifically unsatisfying. An even bigger issue is that galaxies are usually so far apart that one wouldn't expect collisions between them to occur in just 6,000 years. Yet, computer simulations do a very good job of reproducing the appearances of colliding galaxies.[13] This strongly suggests that these collisions are the result of real processes over time and not simply illusions caused by God creating some galaxies very close to one another.

A New Suggestion

Recently, Humphreys proposed a theological option in which God caused *c* in deep space to be much greater on Days 1 through 4 of the creation week. Interested readers may view his explanation online.[14] Although it's too early to unreservedly endorse, this option appeals to me for a couple of reasons.

A much higher deep-space speed of light is a fairly obvious way to get distant starlight to us quickly. And by limiting this higher value of *c* to Days 1 through 4 of the creation week, it avoids some potential problems that plagued previous "faster speed of light" creationist proposals.

But there's an added bonus. Many people are aware that *c* is the "cosmic speed limit," controlling how fast objects can move through space.[15] But they may not be aware that *c* also controls how fast objects travel through *time*.

In relativity theory, time can be thought of as a dimension like length, height, and width. These four dimensions intertwine to form what physicists call spacetime. All objects travel through spacetime at the speed of light *c*, as physicist Lewis Epstein explains:

> Why can't you travel faster than light? THE REASON YOU CAN'T GO FASTER THAN THE SPEED OF LIGHT IS THAT YOU CAN'T GO SLOWER. THERE IS ONLY ONE SPEED [through spacetime]. EVERYTHING, INCLUDING YOU, IS ALWAYS MOVING AT THE SPEED OF LIGHT.[16]

Note that you aren't moving through space at *c* but through spacetime at *c*.[17] Some of that speed *c* is moving you through space, and some of it is moving you through time. Epstein continues:

> If your speed through spacetime could be increased, that increase could be used to increase your speed through time or through space or a combination of the two. But no one knows how to increase it.[16]

No one, that is, except the Lord Jesus! He could have decreed that on Days 1 through 4, *c* in deep space was trillions of times greater than *c* near Earth. Because *c* controls how fast objects move through spacetime, this would automatically imply deep-space objects were moving through spacetime trillions of times faster.

That implies that deep-space clocks were ticking trillions of times faster than clocks on or near Earth. This would give ample time (as measured by clocks "out there") to explain deep-space observations implying long ages while also getting distant starlight to us quickly.

Creation critics may not like the suggestion that *c* in deep space was once much greater than it is now, but since none of us can actually measure *c* in deep space (either now or in the past), this is a philosophical objection, not a scientific one. Within a biblical worldview, it makes perfect sense that the Lord Jesus might have done something like this. After all, He indicated in His Word His intention to get light to Earth quickly (Genesis 1:14–19).

Conclusion

The skeptic can only use distant starlight as an argument against recent creation by making unproven (and likely unprovable) assumptions. He also has to ignore similar difficulties in the Big Bang model. So, despite superficial appearances, distant starlight is really a philosophical objection to recent creation, not a scientific one.[18]

Creation scientists haven't yet nailed down all the details of the answer to this question, but they do offer plausible solutions that agree with known physics as alternatives to dubious evolutionary philosophical assumptions.

References

1. Simple geometry shows that some stars are thousands of light-years from Earth, and these measurements can be used to calibrate still greater distances of millions and billions of light-years.
2. *Cosmos: A Spacetime Odyssey*. 2014. Season 1, Episode 4. "A Sky Full of Ghosts." Brannon Braga. Aired March 30, 2014 on Fox.
3. The Big Bang's horizon problem is that far-apart regions of space have the same temperature for the cosmic microwave background radiation. This could be achieved by radiant energy traveling at the speed of light between these regions. But even 13.8 billion years is insufficient time for light to travel such great distances.

4. Inflation, as originally proposed, is a very brief, faster-than-light expansion of space right after the Big Bang. It solves the horizon problem in that regions of space that are far apart today were, before inflation, sufficiently close together for radiant energy to equalize their temperatures. However, inflation is totally ad hoc and has serious problems of its own.

5. Hebert, J. Big Bang Blowup at *Scientific American*. *Creation Science Update*. Posted on ICR.org May 29, 2017, accessed February 28, 2022.

6. Hebert, J. Big Bang Evidence Retracted. *Creation Science Update*. Posted on ICR.org February 12, 2015, accessed February 28, 2022.

7. Hebert, J. Most Distant Rotating Disk Galaxy Challenges Secular Models. *Creation Science Update*. Posted on ICR.org May 28, 2020, accessed February 28, 2022.

8. Why No One Has Measured the Speed of Light. Veritasium. Posted on youtube.com October 31, 2020, accessed February 28, 2022.

9. Most creationists no longer favor the light-created-in-transit idea. See Hebert, J. 2020. How Can We See Distant Starlight in a Young Universe? In Morris, H. M., III et al. 2020. *Creation Basics & Beyond: An In-Depth Look at Science, Origins, and Evolution*, 2nd ed. Dallas, TX: Institute for Creation Research, 403–409.

10. Humphreys, D. R. 1994. *Starlight and Time: Solving the Puzzle of Distant Starlight in a Young Universe*. Green Forest, AR: Master Books.

11. This does not violate Scripture because the days of creation are clearly measured with respect to clocks on Earth ("and there was evening and morning, the first day," Genesis 1:5, etc.), not by clocks in deep space.

12. Comment by physicist P. W. Dennis at a cosmology panel discussion, August 1, 2018, at the Eighth International Conference on Creationism.

13. Galaxy Collisions: Simulation vs Observations. NASA. Posted on svs.gsfc.nasa.gov September 25, 2015, accessed April 28, 2022.

14. Humphreys, D. R. Toward a More Biblical Cosmology. Gateway Creation Conference. Posted on youtube.com November 22, 2021, accessed April 21, 2022.

15. Objects with mass cannot accelerate up to or beyond *c*.

16. Epstein, L. 2000. *Relativity Visualized*. San Francisco, CA: Insight Press, 78–85. Emphasis in original. Epstein skillfully uses cartoons and diagrams to visually explain the concepts of relativity theory with a minimum of equations. His book is a wonderful resource for both scientists and laypeople.

17. See also theoretical physicist Dr. Sabine Hosenfelder's video "Do we travel through time at the speed of light?" Posted on youtube.com August 29, 2020, accessed March 1, 2022.

18. Dennis, P. W. 2018. Consistent Young Earth Relativistic Cosmology. In *Proceedings of the Eighth International Conference on Creationism*. J. Whitmore, ed. Pittsburgh, PA: Creation Science Foundation, 14–35.

16

"HOT JUPITER" ON THE VERGE OF DESTRUCTION

Jake Hebert, Ph.D.

Astronomers have discovered a Jupiter-size extrasolar planet (exoplanet) orbiting so close to its host star that it is "perilously close" to the distance where it can be torn apart by gravitational forces.[1–3] They call these planets "hot Jupiters" because they are rather large planets with masses comparable to that of Jupiter, and their nearness to their host stars causes them to have high temperatures.

This particular "hot Jupiter," designated as NGTS-10b, is about 1,000 light-years from Earth.[2] Planets with smaller orbits take less time to go around their host stars than do planets with larger orbits. NGTS-10b orbits its host star in just 18 hours! Or to put it another way, the length of its year is less than one Earth day![1,2] This is the shortest orbital period yet observed for an "ultra-short period" hot-Jupiter planet.[2]

Dr. Daniel Bayliss, one of the astronomers making the discovery, said,

> Everything that we know about planet formation tells us that planets and stars form at the

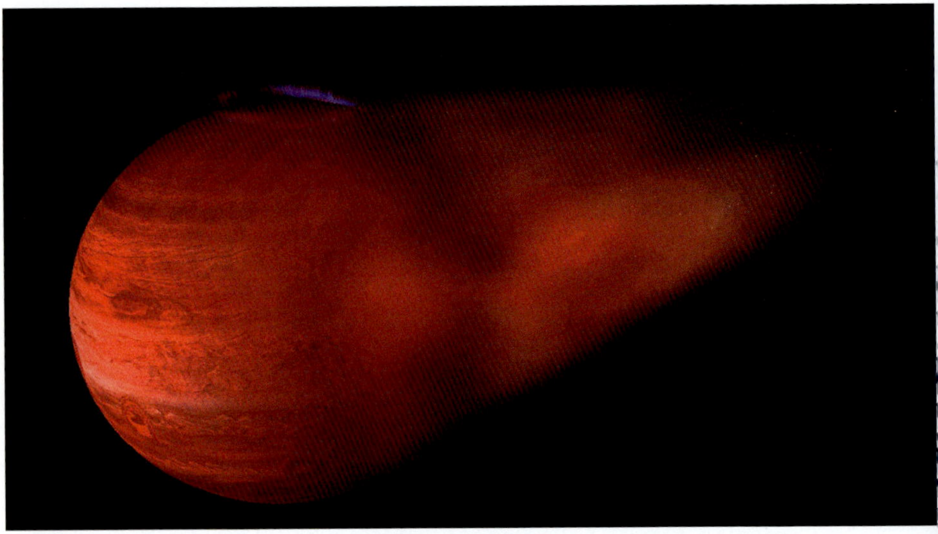

Artist's interpretation of the destruction of a hot-Jupiter planet

> same time. The best model that we've got suggests that the star is about ten billion years old and we'd assume that the planet is too. Either we are seeing it in the last stages of its life, or somehow it's able to live here longer than it should.[2,4]

The suggestion that the planet can "somehow" remain and "live" at this distance from the star "longer than it should" seems odd. Why would astronomers not just simply accept that the planet is in its last stages of its planetary "life"? Why even bring up the possibility that some unknown process could preserve the planet at this destructive distance for a long time?

The reason involves conventional astronomers' belief in billions of years. Scientists estimate that hot Jupiters so close to their parent stars will be destroyed in millions of years. In this particular case, they estimate that NGTS-10b will be destroyed in 38 million years, although there is considerable scatter in this number.[1] This is a long time, but it is only a tiny fraction of the 10 billion years of this planet's supposed lifetime. For this reason, it seems unlikely that astronomers would be "lucky" enough to observe hot Jupiters in their death throes:

> Back in 1996, when the very first hot Jupiters had been discovered, Fred Rasio (Northwestern University) and his colleagues predicted that their orbits could very slowly decay due to tidal interactions with their parent stars. However, the rapid change in WASP-12b's orbital period implies that we're observing the planet at the last moments [here "moment" means a million years!] of its inspiral. That might seem a rather improbable coincidence, but David Latham (Center for Astrophysics, Harvard & Smithsonian) says, "one such case could be the result of a lucky accident."[5]

Perhaps this single case is a lucky accident. However, astronomers have detected several hot Jupiters that apparently have already started their "death spirals" into their host stars.[5–9] This is true even for hot Jupiters having periods longer than a day, such as WASP-12b, mentioned in the above quotation.[5] Some of these planets have undergone measurable decreases in the lengths of their orbital periods.[5,9] One such hot Jupiter is thought to only have 550,000 years left before it is destroyed![8] And there could potentially be many more dying hot Jupiters that we simply don't know about yet. Dr. Bayliss noted:

> We are either very lucky to catch them [hot Jupiters] in this short period orbit, or the processes by which the planet migrates into the star are less efficient than we imagine, in which case it can live in this configuration for a longer period of time."[2]

Yes, perhaps we are "very lucky." And it should be acknowledged that hot Jupiters are relatively easy for astronomers to detect. But if a "selection effect" is a sufficient explanation for the observation of multiple "dying" hot Jupiters so close to their host stars, why would astronomers talk about being "very lucky" to see them? And why would they make the ad hoc suggestion that these planets can "somehow" remain close to their host stars for longer than expected?

Clearly, multiple ultra-short period hot Jupiters this close to their host stars surprised conventional astronomers, hence the ad hoc suggestion that they can "somehow" remain close to their host stars for long ages. Of course, there is another possibility they seem unwilling to consider. Perhaps these hot Jupiters and their host stars simply are much younger than conventional astronomers assume. If our universe is young, it wouldn't be at all surprising to observe multiple hot Jupiters having already begun their death spirals. Their orbital decay, even when measurable, would still cause very little change in these planets' orbits in 6,000 years, even if God originally placed them very close to their current locations.

This is not the first time conventional astronomers have been surprised that they were "lucky" enough to observe changes that should, over billions of years, be very short-lived.[10-12]

Observations of hot Jupiters so close to their host stars make more sense if the stars and planets are young instead of old, just as we would infer from a straightforward reading of Scripture.

References

1. McCormac, J. et al. 2020. NGTS-10b: the shortest period hot Jupiter yet discovered. Monthly *Notices of the Royal Astronomical Society*. 493 (1): 126-140.
2. 18-hour year planet on edge of destruction. Warwick News & Events. University of Warwick. Posted February 20, 2020 on warwick.ac.uk, accessed February 28, 2020.
3. Technically, it is the difference in the strength of gravity on the planet's near side and on its far side that can tear the planet apart.
4. Conventional planet formation theories do not allow for Jupiter-like planets to form close to their host stars, so conventional astronomers must assume these giant planets started out much farther away and then migrated inward.
5. Schilling, G. Planet WASP-12b Might Be on a Death Spiral Into Its Parent Star. *Sky & Telescope*. Posted on skyandtelescope.org September 3, 2019, accessed February 28, 2020.
6. Beatty, K. Watching a Planet's Death Spiral? *Sky & Telescope*. Posted on skyandtelescope.org October 12, 2010, accessed March 9, 2020.
7. Dockrill, P. Astronomers Have Discovered a Potential New Planet Locked in a Death Spiral. ScienceAlert. Posted on sciencealert.com June 10, 2016, accessed March 9, 2020.
8. Rice, D. Distant, massive planet spirals toward a fiery death into its sun. *USA Today*. Posted on usatoday.com March 24, 2017, accessed March 9, 2020.
9. Jones, D. et al. 2018. Evidence of a Decreasing Orbital Period for the Prototype Hot Jupiter System HD 189733: Implications for the Angular Momentum Evolution of Hot Jupiter Systems. *Research Notes of the American Astronomical Society*. 2 (1): 36.
10. Hebert, J. The Hubble 'Pillars of Creation' Revisited. *Creation Science Update*. Posted on ICR.org January 19, 2015, accessed February 28, 2020.
11. Hebert, J. Youthful Solar System Bodies Puzzle Evolutionary Scientists. *Creation Science Update*. Posted on ICR.org February 13, 2013, accessed February 28, 2020.
12. Thomas, B. Saturn's Ring Rain Rates Run Fast. *Creation Science Update*. Posted on ICR.org January 15, 2019, accessed February 28, 2020.

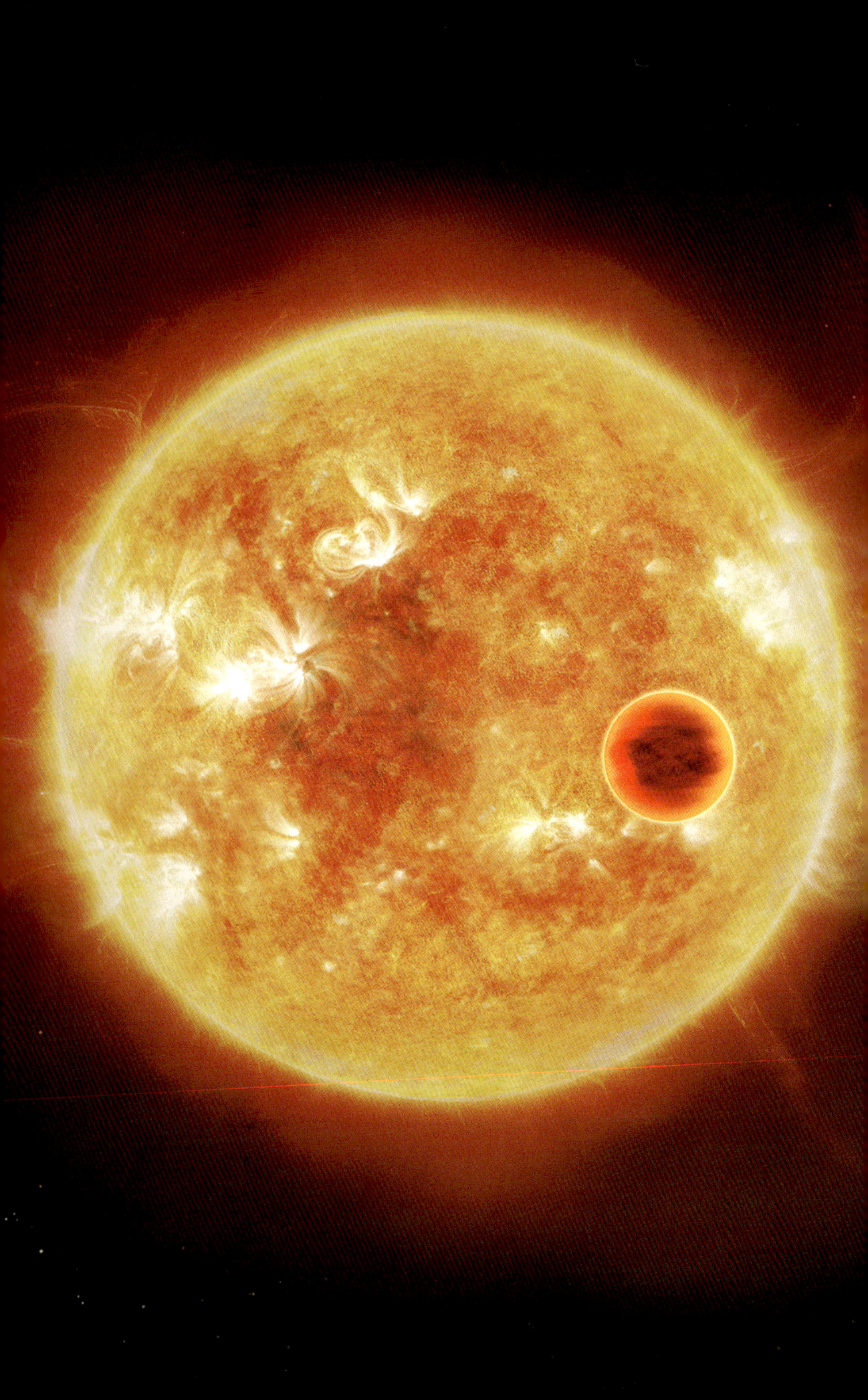

17
NEW EXOPLANET'S YEAR IS JUST 16 HOURS

Jake Hebert, Ph.D.

A team from the Massachusetts Institute of Technology (MIT) used NASA's Transiting Exoplanet Survey Satellite (TESS) to discover a gas giant in another solar system. The planet holds a new record for being a hot, Jupiter-like exoplanet with the shortest year.[1]

Designated TOI-2109b, the planet takes just 16 hours to orbit its host star, while the earlier record-holder orbited its host star in 18 hours.[2] TOI-2109b is called an "ultrahot Jupiter" because it, like Jupiter, is a gas giant, but it orbits so closely to its host star that its dayside is as hot as the surface of a small star! The planet's short orbital period is a consequence of a well-known rule of celestial mechanics called Kepler's Third Law of Planetary Motion. Planets close to their host stars must orbit those stars very quickly.

In fact, astronomers think that TOI-2109b is so close to its star that its orbit is slowly decaying, not unlike bathwater circling a drain. They think they might be able to observe subtle changes in the planet's orbit in just a couple of years.[1]

Do exoplanets in general, and this exoplanet in particular, have any creation implications? Yes. Conventional astronomers think solar systems form from rotating clouds of gas and dust over billions of years. A consequence of this theory is that planets should both rotate on their axes and revolve around their stars in the same way as their host stars. Their orbits should also lie in the same plane (or very close to the plane) as the equator of their host stars. Yet these expectations are routinely contradicted.[3] Of course, since the Lord Jesus created the planets, He could certainly give them any orbital characteristics He desired, not just those that happen to agree with a given conventional hypothesis.

This particular exoplanet also has implications for the age of the universe, as conventional astronomers were very surprised to find hot Jupiters and ultrahot Jupiters that have already begun "death spirals" into their host stars. A 2019 report said that after the first ones were observed, theoretical astrophysicist Fred Rasio "predicted that [hot Jupiter] orbits could very slowly decay due to tidal interactions with their parent stars."[4] However, the rapid orbital decay of a subsequently discovered hot Jupiter "implies that we're observing the planet at the last moments of its inspiral."[4] In context, "the last moments" of exoplanet WASP-12b's inspiral is estimated to be a few million years. This may seem like a long time, but it is a figurative "blink of an eye" for a planet thought to be billions of years old.

Conventional astronomers thought it extremely unlikely they would be able to observe death spirals of more than one hot Jupiter or ultrahot Jupiter. Yet multiple hot Jupiters with decaying orbits have been observed.[2] Another ultrahot gas giant is losing its atmosphere so quickly that even conventional scientists acknowledge that it should no longer have an atmosphere after billions of years. Yet it does have an atmosphere—and a substantial one at that![5]

Exoplanets often seriously challenge conventional planet formation scenarios, and multiple observations of hot gas giants with decaying orbits are surprising if these planetary systems really are billions of years old.

Challenges confront both the young-universe and old-universe positions,[6] but the preponderance of evidence here on Earth and within our solar system strongly favors relative youth.[7] Even in deep space, where evidence of great age often seems strongest, anomalies (including ultrahot gas giants like TOI-2109b improbably teetering on the verge of destruction) indicate that all is not well with the Big Bang model and its 13.8 billion years of assumed history.[8]

References

1. Chu, J. One year on this giant, blistering hot planet is just 16 hours long. *MIT News*. Posted on mit.edu November 23, 2021, accessed November 29, 2021.
2. Hebert, J. "Hot Jupiter" on the Verge of Destruction. *Creation Science Update*. Posted on ICR.org March 19, 2020, accessed November 29, 2021.
3. Musser, G. Why don't exoplanets match astronomers' expectations? A dispatch from the American Astronomical Society Meeting. *Scientific American*. Posted on scientificamerican.com January 13, 2011, accessed November 29, 2021.
4. Schilling, G. Planet WASP-12b Might Be on a Death Spiral Into Its Parent Star. *Sky & Telescope*. Posted on skyandtelescope.org September 3, 2019, accessed November 29, 2021.

5. Hebert, J. Hot Neptune Atmosphere "Shouldn't Exist." *Creation Science Update*. Posted on ICR.org November 5, 2020, accessed November 29, 2021.
6. Hebert, J. 2021. Does the Universe Look Old? *Acts & Facts*. 50 (10): 19.
7. Hebert, J. 2018. Our Young Solar System. *Acts & Facts*. 47 (9): 10–14.
8. Hebert, J. 2019. Deep-Space Objects are Young. *Acts & Facts*. 48 (9): 10–13.

18
HOT-NEPTUNE ATMOSPHERE "SHOULDN'T EXIST"

Jake Hebert, Ph.D.

An exoplanet 260 light-years away is being described as the first of its kind ever detected.[1,2] This exoplanet, catalogued as LTT 9779b, is called an "ultra-hot Neptune" because of its large size and nearness to its host star. This exoplanet is so close that its surface temperature is over 1,700°C, and the length of its year—the time it takes to orbit its star—is just 19 hours! Astronomers think the planet lacks a solid surface and that its atmosphere is substantial, about 9% of the planet's total mass. But the atmosphere is so hot that, by conventional reckoning, *it should not exist*:

> The system itself is around half the age of the sun, at 2 billion years old, and given the intense radiation, a Neptune-like planet would not be expected to keep its atmosphere for so long, providing an intriguing puzzle to solve; how such an improbable system came to be.[1]

Conventional scientists have offered hypotheses to explain LTT 9779b's atmosphere.[1] They suggest it

Artist's interpretation of LTT 9779b

started out as a much larger gas giant and most of its atmosphere was siphoned off by the gravity of its star, leaving behind a much smaller gas planet but one that still retains a significant atmosphere. Or perhaps LTT 9779b did not start out so close to its host star but was nudged relatively recently into its current position via collisions with other bodies. The first option sounds a little more plausible than the second and is predicted (or at least allowed) by models, but these models "require some fine tuning."[1] It's not clear from the news article whether or not the models *themselves* need additional tuning, or whether the first scenario is so unlikely that it can only happen if very special conditions are met (probably the latter). In any case, conventional astronomers were very surprised that a planet this close to a star would have a substantial atmosphere.

There's a very simple solution to this mystery, but it's one that conventional astronomers are loath to consider: What if the true age of LTT 9779b is far, far less than two billion years? What if its age is only about 6,000 years, as the Bible implies? In that case, the planet could still retain a substantial atmosphere despite its nearness to the host star.

Make no mistake about it, the assumption of "deep time" is at the heart of this mystery. Astronomer James Jenkins from the University of Chile stated, "We believe these planets get stripped of their atmospheres *over cosmic time*, ending up as so-called Ultra Short Period planets."[3] Another news report said, "The age of this system is 2 billion years. At this high temperature, the planet's atmosphere should have evaporated long ago, early in the system's life."[4]

Conventional astronomers think stars and planets form naturally from spinning clouds of gas and dust. If this were the case, one would expect all planets to orbit their host stars in planes that coincide, or nearly coincide, with the host star's equator. And they should orbit their stars in the same direction that the stars themselves are rotating. But many exoplanets have highly inclined orbits, and some orbit in the opposite direction as their host stars (i.e., counterclockwise when the host star rotates clockwise, or vice versa). As *Scientific American* said in 2011, "Observers in any field of science take a peculiar pleasure in seeing their theorist colleagues collapse into sobbing heaps, but it happens with unnerving regularity with exoplanets. Modelers have consistently failed to predict the diversity of planetary systems out there."[5]

At ICR.org we've documented many such exoplanet "surprises."[6–8] And some of these exoplanets, such as LTT 9779b, call into question the idea of "billions of years."[9]

Naturalistic theories are inadequate to explain the origin of planets, both those in and those beyond our solar system. Rather, these planets were created by the Lord Jesus Christ on Day 4 of the creation week just a few thousand years ago.[10]

References

1. Thorley, P. The first ultra-hot Neptune, LTT 9779b, is one of nature's improbable planets. *Phys.org*. Posted on phys.org September 22, 2020, accessed October 23, 2020.
2. New study details atmosphere on 'hot Neptune' 260 light years away that 'shouldn't exist.' *Phys.org*. Posted on phys.org October 23, 2020, accessed October 23, 2020.
3. Reference 1, emphasis added.
4. Carr, S. Data reveals evidence of molecular absorption in the atmosphere of a hot Neptune. University of New Mexico news release. Posted on news.unm.edu October 26, 2020, accessed October 26, 2020.
5. Musser, G. Why don't exoplanets match astronomers' expectations? A dispatch from the American Astronomical Society meeting. *Scientific American*. Posted on scientificamerican.com January 13, 2011, accessed October 23, 2020.
6. Thomas, B. Exoplanet Discoveries Demolish Planet Formation Theories. *Creation Science Update*. Posted on ICR.org January 24, 2011, accessed October 23, 2020.
7. Thomas, B. Planet's Reverse Orbit a New Twist in Old Evolutionary Story. *Creation Science Update*. Posted on ICR.org September 18, 2009, accessed October 23, 2020.
8. Hebert, J. Intermediate Gas Giants Challenge Planet Formation Theory. *Creation Science Update*. Posted on ICR.org January 24, 2019, accessed October 23, 2020.
9. Hebert, J. "Hot Jupiter" on the Verge of Destruction. *Creation Science Update*. Posted on ICR.org March 19, 2020, accessed October 23, 2020.
10. Genesis 1:14–19; John 1:1–3; Colossians 1:16.

19
MOST DISTANT ROTATING DISK GALAXY CHALLENGES CONVENTIONAL MODELS

Jake Hebert, Ph.D.

Astronomers have determined that a distant galaxy discovered three years ago is rotating, making it the most distant rotating disk galaxy yet observed.[1] This disk galaxy has been designated DLA0817g but has been nicknamed the "Wolfe Disk" after the late astronomer Arthur M. Wolfe. Astronomers used the Atacama Large Millimeter/submillimeter Array (ALMA), a collection of 66 radio telescopes in northern Chile, to verify that the galaxy is rotating. The research has been published in the journal *Nature*.[2] The press release acknowledges that the Wolfe Disk "provides a challenge for many galaxy formation scenarios."[1]

The astronomers used an indirect method to discover the Wolfe Disk. They realized that light from a quasar more distant than the Wolf Disk had been absorbed by a giant cloud of hydrogen gas. Their discovery of the gas cloud led to their discovery of the disk galaxy within the hydrogen cloud.

Big Bang astronomers assume that light from a

Schematic representation of a rotating disc galaxy. The outer bands of galaxies rotate more slowly than the inner bands.

galaxy billions of light-years away requires billions of years to reach us. While this may seem reasonable, creation scientists question this assumption.[3,4] By Big Bang reckoning, this means that we are seeing these distant objects not as they are now but as they were billions of years ago. Therefore, Big Bang astronomers expect these distant galaxies to be immature, not having time to evolve and develop. However, they have repeatedly been surprised by distant galaxies that are more mature than would be expected by Big Bang reckoning.[5–9] As reported by the ALMA press release,

> In most galaxy formation scenarios, galaxies only start to show a well-formed disk around 6 billion years after the Big Bang. The fact that the astronomers found such a disk galaxy when the Universe was only ten percent of its current age, indicates that other growth processes must have dominated.[1]

Or perhaps it illustrates that something is wrong with conventional ideas about galaxy formation or that something is wrong with the assumption that distant light takes billions of years to reach us or perhaps both.

Distant "early" galaxies can show differences with nearby, more "recent" galaxies, and conventional astronomers attribute those differences to an evolutionary process. However, the study's lead author, Marcel Neeleman of the Max Planck Institute for Astronomy, went on to say,

> "The fact that we found the Wolfe Disk using

Spiral galaxy Caldwell 30

> this [indirect] method, tells us that it belongs to the normal population of galaxies present at early times," said Neeleman. "When our newest observations with ALMA surprisingly showed that it is rotating, we realized that early rotating disk galaxies are not as rare as we thought and that there should be a lot more of them out there."[1]

So, not only are these distant galaxies a problem for the Big Bang, but there may be many more of them than originally thought!

The fact that Big Bang theorists are continually being surprised by their observations is a strong indication that the Big Bang model, despite popular perception, is not a good scientific model.[10,11] Why then are so many Christians apparently intimidated

Spiral galaxy NGC 3147

into accepting it, especially when it contradicts the Genesis history at so many points?[12]

References

1. ALMA Discovers Massive Rotating Disk in Early Universe. ALMA Observatory Press Release. Posted on almaobservatory.org May 20, 2020, accessed May 20, 2020.
2. Neeleman, M., et al. 2020. A cold, massive rotating disk galaxy 1.5 billion years after the Big Bang. *Nature*. 581 (7808): 269–272.
3. Distant starlight in a young universe is used by many as an objection to biblical creation. However, there are subtle, unspoken assumptions beneath this argument, assumptions that may seem reasonable but have

not been shown to be true. Likewise, the Big Bang has its own version of this problem (the horizon problem), and this is one of the main reasons that inflation theory was included in the Big Bang. In fact, the problem of "too mature galaxies in the early universe" discussed in this article can be thought of as *another* light-travel time problem for the Big Bang, since the expectation that very distant galaxies should look immature is based in part on the *assumption* that distant starlight takes billions of years to reach us. See also reference 4.

4. Coppedge, D. F. 2007. The Light-Distance Problem. *Acts & Facts*. 36 (6).
5. Thomas, B. Distant Galaxies Look Too Mature for Big Bang. *Creation Science Update*. Posted on ICR.org November 30, 2011, accessed April 22, 2020.
6. Thomas, B. 'Old' Galaxy Found in 'Young' Part of Universe. *Creation Science Update*. Posted on ICR.org May 24, 2011, accessed April 22, 2020.
7. Thomas, B. Distant Galactic Cluster Should Not Exist. *Creation Science Update*. Posted on ICR.org May 21, 2010, accessed April 22, 2020.
8. Thomas, B. Secrets From the Most Distant Galaxy. *Creation Science Update*. Posted on ICR.org November 18, 2010, accessed April 22, 2020.
9. Hebert, J. 'Early' Spiral Galaxy Surprise. *Creation Science Update*. Posted on ICR.org May 1, 2020, accessed May 21, 2020.
10. Coppedge, D. F. 2008. The Message in Surprise Effects. *Acts & Facts*. 37 (8): 15.
11. Conventional astronomers may object that star and galaxy formation are not actually parts of the Big Bang model *per se*. While *technically* true, this is a cop-out. The Big Bang model purports to explain how our present-day universe came to be, and stars and galaxies are a conspicuous part of that universe. Hence, an inability to explain star and galaxy formation is a problem for the model.
12. Morris, J. D. 1997. Is the Big Bang Biblical? *Acts & Facts*. 26 (5).

20
HUBBLE PHOTOGRAPHS "MOST DISTANT STAR"

Frank Sherwin, D.Sc. (Hon.)

Since its launch 32 years ago, the Hubble Space Telescope (HST) has taken incredible pictures of the universe. With the aid of a phenomenon called gravitational lensing, the Hubble took a remarkable picture of a star that astronomers say is the farthest individual star to date.

Earendel is indicated by an arrow in this image from the Hubble Space Telescope. The star is 28 billion light-years from Earth.

> Earendel is unique because Hubble was able to observe the star independently because of its alignment between earth and a galaxy that provides natural magnification a thousand-fold or more.[1]

Astronomer Brian Welch of Johns Hopkins University nicknamed the star Earendel, which is at least 50 times the mass of our sun and millions of times brighter. Earendel appears to be even older than a 2018 discovery (also set by Hubble) of an enormous blue star nicknamed Icarus.

Evolutionists maintain Earendel emitted its light allegedly within the universe's first billion years (7% of its current age), and its light has taken 12.9 billion years to reach Earth. According to the Big Bang theory, this was when early stars were composed of massive amounts of hydrogen, helium, and trace amounts of lithium. In other words, this might be a legendary Population III star. Welch stated, "Earendel existed so long ago that it may not have had all the same raw materials as the stars around us today."[2] Based on evolutionary reasoning, Earendel should have almost no chemical elements other than hydrogen and helium. Astronomers hope to analyze the star's spectra (information about its temperature, chemical composition, and luminosity) with the James Webb Space Telescope.

But if the star has significant amounts of other chemical elements, this will contradict evolutionary expectations, because the star is also observed alongside "young but mature" galaxies in deep space.[3]

What about the great distances of stars and galaxies? If they are billions of light-years away, then shouldn't their light take billions of years to get here? While this conclusion may seem reasonable, it is an unproven assumption. Even mainstream physicists and astronomers acknowledge that distant starlight can theoretically get here quickly—in fact, in no time at all![4]

Although creationists do not agree with the great ages or deep evolutionary time, they certainly agree with the psalmist who said, "The heavens declare the glory of God; and the firmament shows His handiwork."[5]

References

1. Rae, A. Hubble telescope discovers farthest single star to date. *NBC2 News*. Posted on nbc-2.com April 3, 2022, accessed April 3, 2022. See also: Welch, B. et al. 2022. A highly magnified star at redshift 6.2. *Nature*. 603: 815–818.
2. 28 Billion light-years away: the most distant star ever discovered. *Discovery*. Posted on discovery.com March 31, 2022, accessed April 3, 2022.
3. Hebert, J. Most Distant Rotating Disk Galaxy Challenges Secular Models. *Creation Science Update*. Posted on ICR.org May 28, 2020.
4. For an interesting discussion, watch "Why No One Has Measured The Speed Of Light" by Veritasium, available on YouTube.
5. Psalm 19:1.

21
STARS DEFY BIG BANG

Jake Hebert, Ph.D.

Recent measurements by astronomers at the University of Cologne (Germany) and Masaryk University (Brno, Czech Republic) have shown that a fast-moving star orbits the heart of our Milky Way galaxy in just four years.[1,2] Teasing out this information was an impressive technical feat that required nearly 20 years' worth of data. This fast-moving star, dubbed S4716, and others like it are problematic for evolutionary astronomers.

Astronomers think a supermassive black hole called Sagittarius A* (Sgr A*) is located at the center of our Milky Way galaxy.[3] Some stars orbit very closely to Sgr A*. These have been dubbed "S" stars because of their close association to Sgr A*.[4]

The astronomers were surprised by S4716's short orbit. Lead author and astrophysicist Floyd Peissker said, "For a star to be in a stable orbit so close and fast in the vicinity of a supermassive black hole was completely unexpected and marks the limit that can be observed with traditional telescopes."[1] Astrophysicist Michael Zajaček added,

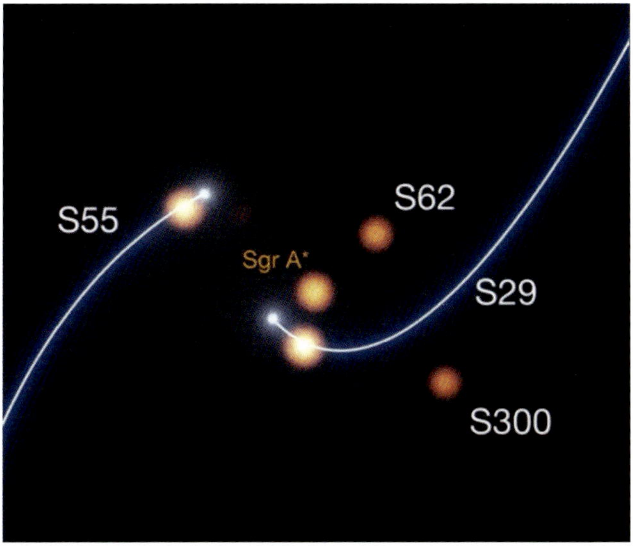

Two stars orbiting Sagittarius A, the black hole at the center of the Milky Way galaxy*

> The short-period, compact orbit of S4716 is quite puzzling....Stars cannot form so easily near the black hole. S4716 had to move inwards, for example by approaching other stars and objects in the S cluster, which caused its orbit to shrink significantly.[1]

If S4716 could not have formed near Sgr A*, it must have formed farther away and then somehow migrated closer to Sgr A*. But based on the comments above, this seems unlikely. The paper's authors speculate on mechanisms that could have made this happen, but they acknowledge that their speculation "needs to be verified in more detailed numerical simulations."[2]

Another star, dubbed S2, or S0-2, orbits the black hole in 16 years and is also problematic for mainstream theorists. Astronomers think it is a young, hot blue "main sequence" star, but they acknowledge "[i]t is challenging to explain the presence of such a young star in close proximity to a supermassive black hole."[5] By evolutionary reckoning, S2 had to have formed less than 10 million years ago, but the strong tidal forces near S2's present location seem to preclude that it could have formed there.[5,6] Could it have formed farther away and then migrated inward? Perhaps, but this process would require a very efficient migration process, and "[c]urrent understanding of the distribution of stars [here], however, does not permit such efficient migration."[5] In other words, the migration process should have taken more than 10 million years, longer than the main sequence lifetime of the star! Simply put, by evolutionary reckoning, S2 should not exist.

Astronomers have suggested other ways to explain S2, but they acknowledge that "[n]one of these possibilities is altogether satisfactory, leaving S0-2 as a paradox of youth in the vicinity of a supermassive black hole."[5]

Of course, this doesn't prove that our Milky Way galaxy is just 6,000 years old, as implied by Scripture, but for Big Bang astronomers, it is one of numerous anomalies and paradoxes that suggest the universe is much younger than 13.8 billion years old.[7–11]

Many creationists think Einstein's theory of relativity is the key to explaining puzzles like distant starlight in a young universe as well as processes in deep space that seem to demand long ages.[12] Relativity theory raises the possibility that clocks in deep space may have ticked much faster than clocks here on Earth. As measured by clocks on Earth, the entire universe is just 6,000 years old, but as measured by clocks in deep space, it could be older. Such a possibility does not contradict Scripture, because the days of creation are clearly measured by clocks here on Earth.[13] But even if this is the means God used to get distant starlight to us quickly, evolutionary age assignments still seem greatly inflated.

Evidence of relative youth is especially strong here on Earth and within our solar system. Yet even in deep space, where evidence of great age seems strongest, there are clues that all is not well with the Big Bang story. This should surprise no one, because our universe is not the result of a big bang but was created by the Lord Jesus Christ just 6,000 years ago.[14,15]

References

1. 8,000 kilometers per second: Star with the shortest orbital period around black hole discovered. Universität zu Köln. Posted on phys.org July 5, 2022, accessed July 5, 2022.
2. Peiβker, F. et al. 2022. Observation of S4716—A Star with a 4 yr Orbit around Sgr A*. *The Astrophysical Journal*. 933 (1).
3. The black hole's name comes from the fact that it is located within the constellation Sagittarius.
4. Yirka, B. The fastest star ever observed. *Phys.org*. Posted on phys.org August 17, 2020, accessed July 7, 2022.
5. Ghenz, A. M. et al. 2003. The First Measurement of Spectral Lines in a Short-Period Star Bound to the Galaxy's Central Black Hole: A Paradox of Youth. *The Astrophysical Journal*. 586: L127–L131.
6. Tidal forces are forces caused by differences in the strength of gravity on one side of an object compared to the other side. Because of their extreme gravity, tidal forces near black holes are very large.
7. Hebert, J. 2018. Our Young Solar System. *Acts & Facts*. 47 (9): 10–14.
8. Hebert, J. 2019. Deep-Space Objects Are Young. *Acts & Facts*. 48 (9): 10–13.

9. Hebert, J. "Hot Jupiter" on the Verge of Destruction. *Creation Science Update*. Posted on ICR.org March 19, 2020, accessed July 7, 2022.
10. Hebert, J. Hot Neptune Atmosphere "Shouldn't Exist." *Creation Science Update*. Posted on ICR.org November 5, 2020, accessed July 7, 2022.
11. Hebert, J. New Exoplanet's Year is Just 16 Hours. *Creation Science Update*. Posted on ICR.org December 31, 2021, accessed July 7, 2022.
12. Hebert, J. 2022. Seeing Distant Starlight in a Young Universe. *Acts & Facts*. 51 (4): 13–15.
13. In Genesis 1, the days of creation are measured by "evening and morning," which clearly implies time measured by Earth clocks.
14. Genesis 1; Psalm 19:1–2; Colossians 1:13–18.
15. Johnson, J. J. S. 2008. How Young is the Earth? Applying Simple Math to Data Provided in Genesis. *Acts & Facts*. 37 (10): 4.

22
THREE HURDLES FOR EVOLUTION

Brian Thomas, Ph.D.

Sure, finch beaks shift their shapes, but does that mean nature invented beaks and birds from stardust? Evolution's proponents often speak as though we witnessed the supposed steps from dust to planets and from single cells to humans. This story attempts to replace the Creator. Since this grand evolutionary narrative gets so much attention in our culture, let's look at three hurdles that only the real Creator could overcome. They put the Lord Jesus back in place as the Author of life and the universe.

The first hurdle concerns the origin of stardust that supposedly turned into us. An anti-Creator bias has basic elements simply coming into existence from nothing. This violates the law of conservation of energy that notes how the total amount of energy in a system stays the same over time. How can a universe conserve the same amount of matter/energy if it goes from nothing to everything?[1]

Maybe matter and energy never had a beginning, making them eternal. This would violate the law of entropy that describes the relentless randomization

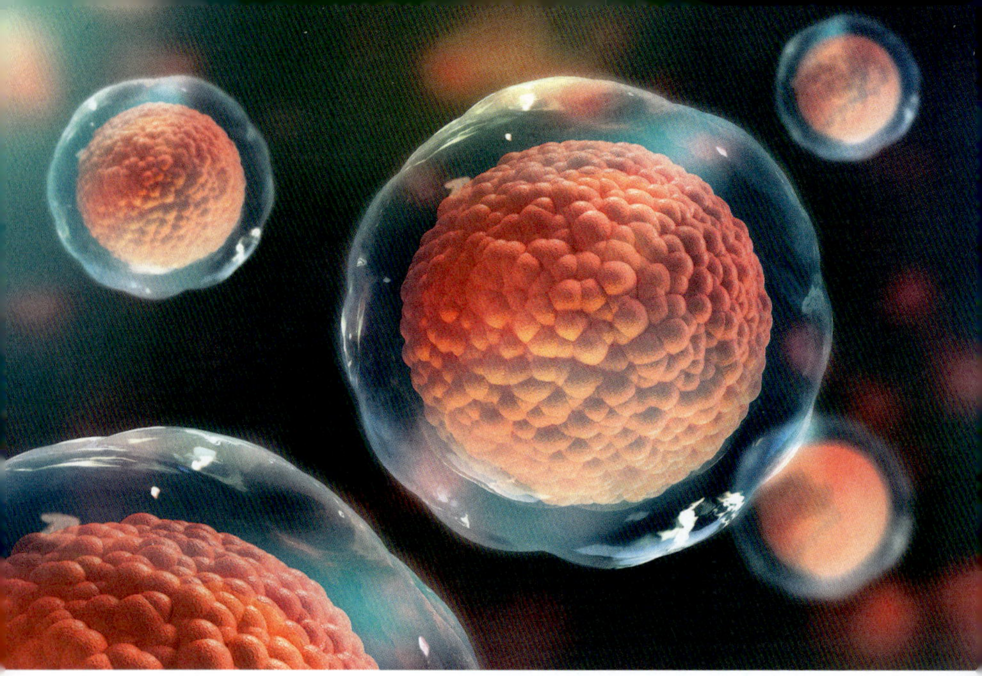

of energy. Since evolution runs into either one hurdle or the other, we have confidence that it takes a Creator to have a creation. And we definitely have a creation. So, when John said about the Lord Jesus that "all things were made through Him" (John 1:3), he offered the best way to leap over the dual hurdles of conservation and entropy—supernatural, instead of natural, origins.

If we pretend that matter and energy did come from nothing, how could they organize themselves into living cells? If dust naturally became cells, then we should see forces make things. Maybe waves should sculpt beach sand into animal sculptures. Maybe wind should put wings onto lizards. But natural forces never add information like specifications or code to matter. From rocks, gases, and whatnot come rocks, gases, and whatnot…only less organized. Even some of the most convinced evolutionists admit that nobody knows how the first cell formed. But without faith that life evolved, one must consider a Creator.

Our last hurdle for evolution concerns the uniquely spiritual qualities of humankind. Both animals and humans can use tools, but only humans can use them to intentionally create beautiful artistic and architectural masterpieces. Both animals and humans can misbehave, but only humans can repent or forgive. Both animals and humans can communicate, but only humans can record communications for future generations. This happened when "holy men of God spoke as they were moved by the Holy Spirit" to write the Bible (2 Peter 1:21). God's Word teaches that Jesus made stars to inhabit the heavens (Genesis 1:15), swimming creatures to inhabit the seas (Genesis 1:21), and people to partake in life with Him forever (John 3:16). Our imagination, moral sensibility, and cross-generational communication all help us relate with our Creator.

With evolution's stories taught as fact so often in our culture, thinking through three of its many hurdles reminds us of just as many reasons why Jesus deserves the credit for our world and why we should know Him.

Reference

1. Some conventional physicists claim a loophole in quantum mechanics allows for universes to pop into existence from nothing. But they forget that quantum mechanics already assumes that space, time, and energy exist.

23
CAN ASTEROIDS SELECT LIFE?

Timothy Clarey, Ph.D.

A 2021 study published in *Science* claimed that an asteroid caused new forms of plants to evolve.[1–3] But can an asteroid really be responsible for selecting new forms of life?

Mónica Carvalho of the Smithsonian Tropical Research Institute, Panama, and her colleagues studied fossil pollen and leaves from Colombia that spanned the K-Pg (Cretaceous-Paleogene) boundary.[1] They concluded that plant diversity declined by 45% at the K-Pg and did not recover for six million years.[1] When plants did recover in the Paleogene, they were markedly different, dominated by flowering plants instead of non-flowering plants.[1]

> "If you returned to the day before the meteorite [asteroid] fall, the forest would have an open canopy with a lot of ferns, many conifers and dinosaurs," said study co-author Carlos Jaramillo. "The forest we have today is the product of one event 66 million years ago."[2]

It sounds like a believable story at first. But impacts can only cause death where they strike. Death cannot create life, let alone new forms of life. Are these scientists assigning intelligence to an asteroid?

Of course, the prevailing conventional story to explain dinosaur extinction involves an asteroid striking the northern edge of the Yucatán Peninsula of Mexico near Chicxulub. But the evidence for the impact is not as compelling as claimed or generally assumed.[4] First, there is not enough melt rock found in the oil wells drilled in and around the crater to justify such a large impact.[4] Second, there is very little iridium found in the wells at the site.[4] (Iridium is a rare earth element that can be produced by an impact and/or by volcanic eruptions.) Third, some of the pre-impact features, like igneous intrusions (dikes), seem unaffected. They should have been obliterated.

In an article for *Answers Research Journal*, I addressed this very issue:

> What is the Chicxulub feature? If it was an impact, it seems to have been much smaller and less significant than generally assumed. Although the data are not conclusive, a case can also be made that there was no impact at the Chicxulub site. Everything can be reasonably explained by volcanic/intrusive activity and the overlap of shallow and deep structural features.[4]

If everything can be explained by something other than a large impact, what is the reason for the sudden change in plant fossils across the K-Pg?

After examining all the evidence, we developed a more satisfying answer. We interpret the vast ma-

The Chicxulub crater area in northwestern Yucatán Peninsula

jority of the sedimentary rock layers as a record of the global Flood that occurred about 4,500 years ago. As the water rose higher and higher, like Genesis 7 describes, more and more land areas became inundated. Evidence shows that the higher elevations in the pre-Flood world apparently had different plants from those in the coastal regions and lowlands.[5] Higher elevations were dominated by mostly flowering plants. In contrast, the lower elevations had mostly non-flowering plants. The K-Pg was simply the boundary between two different ecological zones segregated by elevation differences.

As the global Flood inundated the coastal and lowland regions, non-flowering plants and the dinosaurs were buried in the same strata. This may have occurred between Day 40 and Day 140 of the Flood year.[5] As the water continued to rise, reaching its highest peak on Day 150, it flooded the uplands that mostly contained flowering plants and many of the larger mammals.[5] This buried the large mammals and the flowering plant fossils on top of the earlier-deposited dinosaur-rich layers. Therefore, the Flood explains exactly why the plants change and the dinosaurs disappeared at about the same time. It was the systematic burial of two completely different pre-Flood ecological zones.[5]

Nonetheless, the Chicxulub impact has become the iconic tale that is readily accepted by most of conventional science to explain what seems to be a K-Pg extinction event that appears to have wiped out the dinosaurs and changed the plants, too.[1] Even though there are alternative ways to explain the evidence, they attribute these events to an asteroid because they have no other recourse. Conventional science categorically rejects Earth history as described in the Bible, including the reality of the global Flood.

References

1. Carvalho, M. R. et al. 2021. Extinction at the end-Cretaceous and the origin of modern Neotropical rainforests. *Science*. 372 (6537): 63–68.
2. Crane, L. Asteroid that killed the dinosaurs gave birth to the Amazon rainforest. *New Scientist*. Posted on newscientist.com April 1, 2021, accessed April 5, 2021.
3. Jacobs, B. F. and E. D. Currano. 2021. The impactful origin of neotropical plants. *Science*. 372 (6537): 28–29.
4. Clarey, T. L. 2017. Do the data support a large meteorite impact at Chicxulub? *Answers Research Journal*. 10: 71–88.
5. Clarey, T. 2020. *Carved in Stone: Geological Evidence of the Worldwide Flood*. Dallas, TX: Institute for Creation Research.

24
EARTH REMAINS THE ONLY GOLDILOCKS PLANET

Brian Thomas, Ph.D.

San Francisco State University astronomer Stephen Kane hunts for signs of life in outer space. His team measured light output from a distant star called Wolf 1061.[1] Exoplanets (planets outside our solar system) orbit Wolf 1061, and one of them seemed to promise the presence of liquid water, a prerequisite for life.

Kane's group tested whether the "c" planet of Wolf 1061, called Wolf 1061c, would permit liquid water. The "Goldilocks zone," or habitable zone (HZ), describes the distance from each particular star at which water on an orbiting planet would retain liquid. Life needs liquid water to facilitate all its necessary biochemical reactions.

But excitement dwindled when they overlaid the likely orbital track of planet Wolf 1061c onto the Goldilocks zone. They discovered that the planet orbits so close to its star that any water there would likely cause a runaway greenhouse effect. It would be a hellish sauna, like Venus. Basically, they just crossed

Artist's impression of the planetary system around Wolf 1061

Wolf 1061c off the list of exoplanets that might have liquid water. The final sentence in their technical report reads, "We thus conclude that the system is unlikely to host planets with surface liquid water."[2]

Kane said to SF State News, "The Wolf 1061 system is important because it is so close and that gives other opportunities to do follow-up studies to see if it does indeed have life." Why was he so optimistic after admitting these negative results?

Perhaps he meant to say that the Wolf 1061 system *might have been* important because it is close enough to allow follow-up studies that look for signs of life. Now it seems to be another dead end like everywhere but Earth, so far as we know.

Of the hundreds of exoplanets targeted for faint

hints of distant life, so far none meet the most basic requirement: liquid water. However, if some astronomers ever do verify liquid water on a distant planet, they will almost certainly take it as a sign of life even though water destroys the very chemicals that make cells operate. Water facilitates the right chemistry *inside* already-living cells but terminates those same chemicals when exposed *outside* the protective membrane of a cell.[3]

With Wolf 1061c too close to its star to reasonably expect liquid water and water's power to destroy biochemistry, the idea of life somewhere out there goes from a hope *without* reason to a hope *against* reason.

References

1. Oppenheim, J. SF State astronomer searches for signs of life on Wolf 1061 exoplanet. *SF State News*. Posted on news.sfsu.edu January 13, 2017, accessed January 20, 2017.
2. Kane, S. R. et al. Characterization of the Wolf 1061 Planetary System. arXiv. Posted on arxiv.org December 29, 2016, accessed January 20, 2017.
3. See www.ICR.org/creation-cells.

25
NASA ANNOUNCES 5,000 EXOPLANETS DISCOVERED

Frank Sherwin, D.Sc. (Hon.)

Evolutionary scientists would like nothing better than to find life in any form outside of Earth, but so far this has been lacking. The pages of the book on astrobiology (the study of life in the universe) remain blank, figuratively speaking. Regardless, some astronomers have predicted alien contact in 2025,[1] and half of humanity believes in aliens.[2]

In 2022, NASA achieved a space exploration milestone when it confirmed discovery of over 5,000 exoplanets (planets orbiting stars other than our sun).[3] Could there be millions or even trillions of more such planets?

"Surely," some might say, "with all those exoplanets there will be *some* harboring life." No, not really. Fervently believing that life exists on other planets is based on the unscientific assumption that organic (carbon-based) life could spring—through time—from inorganic non-life. Such a process is called abiogenesis (the origin of life) and has never been documented.

One of the co-discoverers of the DNA molecule was Sir Francis Crick of England. He was a brilliant scientist and avowed atheist. Crick knew the futility of abiogenesis and looked beyond Earth for life's origin. He was driven, along with British chemist Leslie Orgel, to put together a bizarre hypothesis that life on earth could have sprung from interstellar seeds! Such unknown seeds would have been spread from an unknown spacecraft sent by an unknown distant civilization several billion years ago. This ironically is called directed panspermia, and it is entirely theoretical; there is no scientific evidence that this occurred.

In November 2013, Tanya Lewis of *Space.com* wrote, "Habitable Earth-Size Planets Common Across the Universe, Study Suggests."[4] But six years later, another writer stated, "a new study...dramatically reduces the number of worlds where scientists will have the best luck finding ET."[5]

Indeed, scientists listened for aliens among the billions of stars and came up empty-handed.

> The array used to conduct the search is made up of 4,000 antennas in the Western Australian desert. The researchers listened in for any kind of alien "technosignatures", or any kind of sign of alien life out there. Because the array can pick up low-frequency radio waves, they hoped it could give some kind of evidence of intelligent life on other worlds. During the seven hours of listening, the group focused on the Galactic Center. The center has been a prime target for astronomers in the past

> because of all the stars it houses. More stars mean more chances for worlds capable of living and intelligent creatures. However, after 7 hours of listening in on the Galactic Center, the researchers turned up absolutely nothing.[6]

Life as we know it requires liquid water, but a tiny minority of these 5,000+ exoplanets have water or water vapor.

> Consider the conditions necessary for a large, permanent ocean of liquid water to form on a planet. An ocean world must move in a nearby circular orbit around a stable star. The distance of the planet from the star must be just right to provide a temperate environment in which water is liquid. Unlike most stars, a water planet's sun must not be a double or multiple star, or the orbital year would have irregular periods of intense heat and cold. The material that accreted [came or was brought together under the influence of gravitation] to form the planet must have included both water and substances capable of forming a solid crust.[7]

But just having water may not be as significant a factor as once thought. Many are starting to argue that the standard definition of habitability—having liquid water on a planet's surface—is not the factor that should guide exoplanet exploration. Instead, the scientists say the field should focus on the chances of detecting alien life, should it exist. "'Planets can be habitable and not have life with any impact,' [Steve Desch, an astrophysicist at Arizona State University,] told researchers at the meeting."[8]

Exoplanet research is, of course, fascinating, but whether they have all the many parameters needed to support organic life is unknown and unlikely. There is no evidence in either Scripture or science that life—especially intelligent life—exists on these exoplanets.[9]

> The heaven, even the heavens, are the LORD's;
> But the earth He has given to the children of men.[10]

References

1. Sherwin, F. 2004. 21 Years Until Alien Contact? *Acts & Facts*. 33 (3).
2. Thomas, B. Poll: Half of Humanity Believes in Aliens. *Creation Science Update*. Posted on ICR.org December 26, 2017.
3. Seo, H. NASA's official exoplanet tally has passed 5,000 worlds. *Popular Science*. Posted on popsci.com March 22, 2022, accessed March 24, 2022.
4. Lewis, T. Habitable Earth-Size Planets Common Across the Universe, Study Suggests. *Space.com*. Posted on space.com November 4, 2013, accessed March 22, 2022.
5. Geggel, L. Alien Life Could Be Hiding Out on Far Fewer Planets Than We Thought. *Live Science*. Posted on livescience.com June 10, 2019, accessed March 22, 2022.
6. Hawkins, J. Scientists searched for aliens among the billions of stars, and the results will surprise you. Yahoo! Finance. Posted on finance.yahoo.com February 24, 2022, accessed March 21, 2022.
7. Garrison, T. and R. Ellis. 2016. *Oceanography*, 9th ed. Boston, MA: Brooks/Cole, 17. This is all the more unique because "according to models of Solar System formation, Earth, as an inner Solar System planet, should have little to no water." Peslier, A. 2020. The origins of water. *Science*. 369 (6507): 1058.
8. Witze, A. Exoplanet Hunter Rethinks Search for Alien Life. *Scientific American*. Posted on scientificamerican.com November 21, 2017.
9. Regarding a Christian response to aliens and UFOs, see Creation Ministries International's program "Aliens and UFOs--What are they really?" posted on creation.com.
10. Psalm 115:16.

26
PREPARE FOR THE BIG NON-EVENT

Frank Sherwin, D.Sc. (Hon.)

In 1950, the famed Italian physicist Enrico Fermi asked his coworkers at the lunch table the simple metaphysical question about possible aliens in the universe, "Where is everybody?" Since then, his question has become the Fermi Paradox: the supposed high estimates for intelligent extraterrestrial life conflicting with cold, interstellar silence.

In 1984, the ambitious Search for Extra Terres-

Allen Telescope Array

Enrico Fermi

trial Intelligence (SETI) was launched to intercept electromagnetic radiation from possible civilizations on other planets. But interest has since waned, and funding has been a problem.[1]

But now there's Breakthrough Listen, a project with $100 million in funding to listen for intelligent extraterrestrial communications. This mission uses thousands of hours on telescopes in facilities throughout the world in an attempt to discover technosignatures (indicators of technology) and biosignatures (biological processes). However,

> No unambiguous technosignatures have been seen to date, but the chances of detection are higher than they have ever been, in large part due to Breakthrough Listen—the most sensitive, comprehensive, and intensive search

> for advanced life on other worlds ever performed.[2]

Two evolutionists are convinced that through the alleged billions of years of the universe's existence, evolution has produced intelligent life somewhere and that we may make contact with aliens in 2025.[3]

In 2011, Howard Smith, a senior astrophysicist at Harvard, stated after surveying 500 planets that we are alone in the universe.[4] Life—intelligent or otherwise—hasn't been found. Nevertheless, predictions continue to be made.[5]

Regardless of hearing nothing from the depths of space, scientists are insisting Earth should get prepared to answer interstellar messages.

> Dr John Elliott, honorary research fellow... said: "Will we ever get a message from ET? We don't know. We also don't know when this is going to happen. But we do know that we cannot afford to be ill prepared—scientifically, socially, and politically rudderless—for an event that could turn into reality as early as tomorrow and which we cannot afford to mismanage. We need to coordinate our expert knowledge not only for assessing the evidence but also for considering the human social response, as our understanding progresses and what we know and what we don't know is communicated. And the time to do this is now."[6]

Evolutionists continue to place their faith in evolutionary processes that will produce—they hope[7]—

intelligent, non-violent beings that will soon contact us.

The Bible is historically accurate. It gives us clear insight into the origin of the universe and Earth. It doesn't mention intelligent extraterrestrials.

References

1. Thomas, B. SETI Funding Linked to Belief in Evolution. *Creation Science Update*. Posted on ICR.org May 9, 2011, accessed November 3, 2022.
2. Science Writer. Breakthrough Listen Collaborates With TESS. *SETI Institute*. Posted on seti.org October 23, 2019, accessed November 4, 2022.
3. Sherwin, F. 21 Years Until Alien Contact? *Creation Science Update*. Posted on ICR.org March 1, 2004, accessed November 4, 2022.
4. Blake, H. Alien life deemed impossible by analysis of 500 planets. *The Telegraph*. Posted on telegraph.co.uk January 23, 2011, accessed November 4, 2022.
5. Wall, M. Signs of Alien Life Will Be Found by 2025, NASA's Chief Scientist Predicts. *Space.com*. Posted on space.com April 7, 2015, accessed November 3, 2022.
6. Knapton, S. Prepare for alien encounter now before it's too late, warn scientists. *The Telegraph*. Posted on telegraph.co.uk November 2, 2022, accessed November 4, 2022.
7. Oluseyi, H. Intelligent life probably exists on distant planets—even if we can't make contact, astrophysicist says. *The Washington Post*. Posted on washingtonpost.org June 19, 2021, accessed November 4, 2022.

CONTRIBUTORS

Dr. Timothy Clarey is the director of research at the Institute for Creation Research and earned his doctorate in geology from Western Michigan University.

Dr. Jake Hebert is a research scientist at the Institute for Creation Research and earned his Ph.D. in physics from the University of Texas at Dallas.

Dr. Frank Sherwin is a science news writer at the Institute for Creation Research. He earned an M.A. in zoology from the University of Northern Colorado and received an Honorary Doctorate of Science from Pensacola Christian College.

Dr. Brian Thomas is a research scientist at the Institute for Creation Research and earned his Ph.D. in paleobiochemistry from the University of Liverpool.

Dr. Jeffrey P. Tomkins is a research scientist at the Institute for Creation Research and earned his Ph.D. in genetics from Clemson University.

IMAGE CREDITS

BigstockPhoto: 8, 42, 53, 82, 102, 142, 144, 152
Brewbooks via Wikimedia Commons: 157
Brews ohare: 104
Pablo Carlos Budassi via Wikimedia Commons: 93, 99
Illustris Collaboration via Wikimedia Commons: 78
iStockPhoto: 31
ESA/ATGmedialab via Wikimedia Commons: 118
ESO via Wikimedia Commons: 22
ESO L.Calçada CC BY 4.0 DEED: 128
ESO/GRAVITY collaboration/Luis Calçada via Wikimedia Commons: 137
Jake Hebert: 30, 88
NASA: 36, 87, 90, 124, 130, 147
NASA/CXC/U.Texas: 19
NASA/ESA: 59, 108, 131
NASA/ESA/CSA: 64
NASA/ESA/CSA/STScI: 56
NASA/ESA/CSA/STScI/Joseph DePasquale, STScI/Anton M. Koekemoer, STScI/
Alyssa Pagan, STScI: cover, 1, 3
NASA/ESA/Brian Welch, JHU/Dan Coe, STScI/Ayssa Pagan, STScI: 133
NASA/GSFC via Wikimedia Commons: 27
NASA/GSFC/CIL: Adriana Manrique Gutierrez via Wikimedia Commons: 48
NASA/JPL/Caltech: 150
NASA/WMAP Science Team: 68, 74
Nolesh via Wikimedia Commons: 114
Donald Pelletier via Wikimedia Commons: 137
Public domain: 84
Quantum Doughnut via Wikimedia Commons: 70
Smithsonian Institution: 158
Georg Wiora, CC BY-SA 3.0. Derivative work Kes47: 57

ATGmedialab—ATG Europe Medialab
Caltech—California Institute of Technology
CXC—Chandra X-Ray Center
CIL—Conceptual Image Laboratory
CSA—Canadian Space Agency
ESA—European Space Agency
ESO—European Southern Observatory
GSFC—Goddard Space Flight Center
JHU—Johns Hopkins University
JPL—Jet Propulsion Laboratory
NASA—National Aeronautics and Space Administration
STScI—Space Telescope Science Institute
U.Texas—University of Texas
WMAP—Wilkinson Microwave Anisotropy Probe

THIS BOOK WAS ADAPTED FROM THE FOLLOWING MATERIALS

Tomkins, J. P. 2023. Creation *Ex Nihilo* Through Jesus Christ. *Acts & Facts*. 52 (1): 14–16.

Thomas, B. A Supernova and the Scripture. *Creation Science Update*. Posted on ICR.org October 28, 2020.

Hebert, J. First Ever Photo of a Black Hole. *Creation Science Update*. Posted on ICR.org April 23, 2019.

Hebert, J. 2019. Continuing Troubles for the Big Bang Model. *Acts & Facts*. 48 (11): 10–13.

Hebert, J. Big Bang Hubble Contradiction Confirmed. *Creation Science Update*. Posted on ICR.org May 16, 2019.

Hebert, J. *New York Times* Editorial: Big Bang Unraveling? *Creation Science Update*. Posted on ICR.org September 14, 2023.

Hebert, J. Webb Telescope Continues to Challenge Big Bang. *Creation Science Update*. Posted on ICR.org January 26, 2023.

Hebert, J. *Astronomy* Magazine: Big Bang in Crisis? *Creation Science Update*. Posted on ICR.org May 21, 2020.

Hebert, J. 2022. James Webb Telescope vs. the Big Bang. *Acts & Facts*. 51 (11): 14–17.

Hebert, J. 2018. Does the Cosmic Microwave Background Confirm the Big Bang? *Acts & Facts*. 47 (6): 10–14.

Hebert, J. Dark Matter Search Keeps Coming Up Empty. *Creation Science Update*. Posted on ICR.org April 9, 2020.

Hebert, J. 2021. Does the Universe Look Old? *Acts & Facts*. 50 (10): 19.

Hebert, J. 2019. Deep-Space Objects Are Young. *Acts & Facts*. 48 (9): 10–13.

Hebert, J. "Early" Spiral Galaxy Surprise. *Creation Science Update*. Posted on ICR.org May 1, 2020.

Hebert, J. 2022. Seeing Distant Starlight in a Young Universe. *Acts & Facts*. 51 (4): 13–15.

Hebert, J. "Hot Jupiter" on the Verge of Destruction. *Creation Science Update*. Posted on ICR.org March 19, 2020.

Hebert, J. New Exoplanet's Year Is Just 16 Hours. *Creation Science Update*. Posted on ICR.org December 13, 2021.

Hebert, J. Hot Neptune Atmosphere "Shouldn't Exist." *Creation Science Update*. Posted on ICR.org November 5, 2020.

Hebert, J. Most Distant Rotating Disk Galaxy Challenges Secular Models. *Creation Science Update.* Posted on ICR.org May 28, 2020.

Sherwin, F. Hubble Photographs "Most Distant Star." *Creation Science Update.* Posted on ICR.org April 11, 2022.

Hebert, J. Stars Defy Big Bang. *Creation Science Update.* Posted on ICR.org July 14, 2022.

Thomas, B. 2021. Three Hurdles for Evolution. *Acts & Facts.* 50 (9): 13.

Clarey, T. Can Asteroids Select Life? *Creation Science Update.* Posted on ICR.org April 29, 2021.

Thomas, B. Earth Remains the Only Goldilocks Planet. *Creation Science Update.* Posted on ICR.org February 16, 2017.

Sherwin, F. NASA Announces: 5,000 Exoplanets Discovered. *Creation Science Update.* Posted on ICR.org March 31, 2022.

Sherwin, F. Prepare for the Big Non-event! *Creation Science Update.* Posted on ICR.org November 14, 2022.

ABOUT THE INSTITUTE FOR CREATION RESEARCH

At the Institute for Creation Research, we want you to know God's Word can be trusted with everything it speaks about—from how and why we were made, to how the universe was formed, to how we can know Jesus Christ and receive all He has planned for us.

That's why ICR scientists have spent more than 50 years researching scientific evidence that refutes evolutionary philosophy and confirms the Bible's account of a recent and special creation. We regularly receive testimonies from around the world about how ICR's cutting-edge work has impacted thousands of people with Christ's creation truth.

HOW CAN ICR HELP YOU?

You'll find faith-building science articles in *Acts & Facts*, our bimonthly science news magazine, and spiritual insight and encouragement from *Days of Praise*, our quarterly devotional booklet. Sign up for FREE at **ICR.org/subscriptions**.

Our radio programs, podcasts, online videos, and wide range of social media offerings will keep you up-to-date on the latest creation news and announcements. Get connected at **ICR.org**.

We offer creation science books, DVDs, and other resources for every age and stage at **ICR.org/store**.

Learn how you can attend or host a biblical creation event at **ICR.org/events**.

Discover how science confirms the Bible at our Dallas museum, the ICR Discovery Center. Plan your visit at **ICRdiscoverycenter.org**.

ICR INSTITUTE FOR CREATION RESEARCH

P. O. Box 59029
Dallas, TX 75229
800.337.0375
ICR.org

ICR'S MISSION STATEMENT

ICR EXISTS TO SUPPORT THE LOCAL CHURCH THROUGH...

WORSHIP

- Glorify Jesus Christ by emphasizing in all ICR resources the credit He is due as Creator.
- Oppose the deification of nature by exposing Darwinian selectionism as an idolatrous worldview.

EDIFICATION

- Help pastors lead, feed, and defend their flocks by providing scientific responses to secular attacks on the authority and authenticity of God's Word.
- Change Christians' view of biology by constructing an organism-focused theory of biological design that highlights Jesus' work as Creator.

EVANGELISM

- Defend the gospel by showing how natural processes cannot explain the miracles in the Bible.
- Counter objections to the gospel by equipping believers with Scripture-affirming science.

RESOURCES FROM ICR

Find out more about these books and other resources at **ICR.org/store**

ICR
INSTITUTE FOR CREATION RESEARCH